Cracking the Case of ISO 9001:2015 for Manufacturing

Also available from ASQ Quality Press:

Cracking the Case of ISO 9001:2015 for Manufacturing

A Simple Guide to Implementing Quality Management in Manufacturing

Third Edition

Charles A. Cianfrani
and John E. (Jack) West

ASQ Quality Press
Milwaukee, Wisconsin

American Society for Quality, Quality Press, Milwaukee 53203
© 2017 by ASQ
All rights reserved. Published 2016
Printed in the United States of America
22 21 20 19 18 17 16   5 4 3 2 1

Library of Congress Cataloging-in-Publication Data

Names: Cianfrani, Charles A., author. | West, Jack, 1944- author.
Title: Cracking the case of ISO 9001:2015 for manufacturing : a simple guide
   to implementing quality management in manufacturing / Charles A.
   Cianfrani, and John E. (Jack) West.
Description: Third edition. | Milwaukee, Wisconsin : ASQ Quality Press, 2017.
   | Includes index.
Identifiers: LCCN 2016032060 | ISBN 9780873899079 (soft cover : alk. paper)
Subjects: LCSH: Quality control—Standards. | ISO 9001 Standard.
Classification: LCC TS156.17.I86 C53 2017 | DDC 658.5/620218—dc23
LC record available at https://lccn.loc.gov/2016032060

ISBN: 978-0-87389-907-9 (Spiralbound)

ISBN: 978-0-87389-907-9 (paperback)

ASQ Mission: The American Society for Quality advances individual, organizational,
and community excellence worldwide through learning, quality improvement, and
knowledge exchange.

Attention Bookstores, Wholesalers, Schools, and Corporations: ASQ Quality Press
books, video, audio, and software are available at quantity discounts with bulk
purchases for business, educational, or instructional use. For information, please
contact ASQ Quality Press at 800-248-1946, or write to ASQ Quality Press, P.O. Box
3005, Milwaukee, WI 53201-3005.

To place orders or to request ASQ membership information, call 800-248-1946. Visit
our website at http://www.asq.org/quality-press.

Quality Press
600 N. Plankinton Ave.
Milwaukee, WI 53203-2914
Email: books@asq.org

ASQ   Excellence Through Quality™

Table of Contents

List of Figures and Tables

Chapter 1

Why Do ISO 9001:2015?

ISO 9001:2015, like its predecessor editions since 1987, continues to provide a model for quality management. While no such model is perfect, this one has proven to be applicable to virtually all sizes of organizations, in every marketplace, and for every product and service category throughout the world for over 25 years.

Why has ISO 9001 become the world's most used standard? Why has it achieved such widespread acceptance and use? Certainly not because of the elegance of the text in the standard, for this writing is among the dullest, most boring prose the human mind and hand have ever crafted. The ISO 9001 standard has survived and flourished because it adds value to how organizations are managed, from the viewpoint of both managers and workers.

Workers like ISO 9001 because it makes life simpler. In an ISO 9001 system, workers have:

- A better understanding of what to do and how to do it

- The ability to ensure that their work meets requirements

- The ability to adjust processes when results are not meeting requirements

- A means to get help in solving problems

- Increased opportunities to communicate problems in a nonthreatening manner by focusing on process issues

- An environment where they are not blamed for issues that can be resolved only by managers

Middle managers have embraced ISO 9001 because it has contributed to better control of processes and resulted in a higher level of consistency throughout the organization. Middle managers find that ISO 9001 has:

- Promoted using facts and data to manage rather than opinions

- Enhanced communication throughout the organization (between management and workers, between departments, and with executive management)

- Encouraged clarity of responsibility and accountability

- Standardized the way things are done, reducing variability and making it easier to solve problems

- Fostered continual improvement as an institutionalized core value, and provided a platform for moving to performance excellence

Top managers find that adopting a formal ISO 9001 quality management system (QMS) helps the organization focus on meeting objectives. Top managers find that ISO 9001 has:

- Improved their organization's ability to fully understand and meet customer requirements in a consistent manner

- Brought greater clarity to the goals and objectives of the organization

- Helped align all employees and processes in the efforts to meet objectives

- Improved bottom-line performance by enhancing revenue and reducing costs, created a competitive advantage in some markets, and enabled their organizations to compete in markets where most other potential suppliers are registered

Perhaps the most important reason for doing ISO 9001 is *survival*. In these times of uncertainty, people are concerned about protecting their jobs.

All of these reasons for embracing ISO 9001:2015 can be reduced to a single effect: it facilitates quality improvement! And quality improvement has two positive impacts on the organization:

- Better processes and reduced variation in production can yield dramatic reductions in cost.

- Better customer satisfaction can yield more sales or repeat business.

If an organization does not have a foundation of uncompromising integrity, adventures into the world of performance excellence or attempts to sustain improvement programs are futile exercises. Compliance with the requirements of ISO 9001 is an essential element of a foundation upon which a successful organization can be built. The requirements may not be sexy or exciting, but unless they are adhered to consistently and well, an organization will not prosper, and may not survive in the contemporary marketplace.

In Chapter 5 of this guide we have included a description of 14 quality tools that may be helpful to you when you are structuring or deploying processes to effectively comply with ISO 9001:2015 requirements. For each tool we describe (1) what is it? (2) where is it used? (3) how is it done? and (4) cautions to be considered when using the tool.

Throughout Chapter 4 of this guide, which addresses each clause and requirement of ISO

9001:2015, you will notice icons with a number next to some headings. The icons and number are reminders for you to consider the applicability of the tool described in Chapter 5 for achieving effective conformity with the specific clause or sub-clause.

This guide is intended to help everyone in an organization participate in creating and sustaining a foundation of integrity, meet requirements and customer expectations, and support robust processes, to the advantage of everyone in the organization and to each of its customers.

Charlie Cianfrani
cianfranic@aol.com

Jack West
jwest92144@aol.com

Chapter 2

Key Differences with Past ISO 9001 Editions

Before starting our journey to understand the specifics and details of the ISO 9001:2015 requirements and how to comply, let's quickly consider content that may appear to be new to many users of the standard, recognizing that what is considered new will vary from organization to organization. This overview will heighten your attention to these areas when we explore the specifics in the more detailed discussion of each clause.

Our list of content that may be considered new to many readers includes:

- Annex SL structure

- Understanding the organization and its context

- Understanding the needs and expectations of interested parties

- Actions to address risks and opportunities

- Organizational knowledge

Details on each of these elements of the standard are presented in other chapters of this book. Here, we include a few high-level comments to provide you with background that may enhance your understanding of the intent of this new content:

- *Annex SL structure.* In ISO 9001:2008 the requirements are contained in clauses 4–8. In ISO 9001:2015 the requirements are contained in clauses 4–10. The new structure of the requirements clauses was dictated by the ISO directives and will be used by all ISO management system standards (MSSs). The intent expressed by ISO leaders (which is arguable) is that a common structure for MSSs will facilitate compliance by organizations.

Some will view the new structure as "new." Others will shrug their shoulders and comment that this new structure is nothing more than "reshuffling the deck." After you work with ISO 9001:2015, you can decide whether the new structure makes sense and adds any value. Your authors have their own opinion, which will remain unstated.

- *Understanding the organization and its context.* These words are new to ISO 9001 and are an attempt to require the organization to consider its strategic direction so it can determine issues of real or potential impact and plan and deploy processes and controls to manage such issues. Although the words may be new, many organizations already have strategic and tactical planning processes in place to address such a requirement. If such processes do not exist, ISO 9001:2015 provides an incentive to consider their implementation. To provide a frame of reference for this concept, Figure 2.1 provides an overview of the processes of the QMS and the extent of what may be included in the appraisal of the context of the organization. This figure is a little different than the figure in the standard, and we believe it is easier to understand.

Figure 2.2 indicates that consideration of the context of the organization should include both internal and external opportunities and threats. Such thinking may be common in many organizations but new to others.

• *Understanding the needs and expectations of interested parties*. Interested parties? What does this mean? In prior editions of ISO 9001 there was strong concern for meeting customer, statutory, and regulatory requirements, but no reference to any other parties. However, given the requirement related to the organization and its context, there is now a need to consider other relevant interested parties in addition to

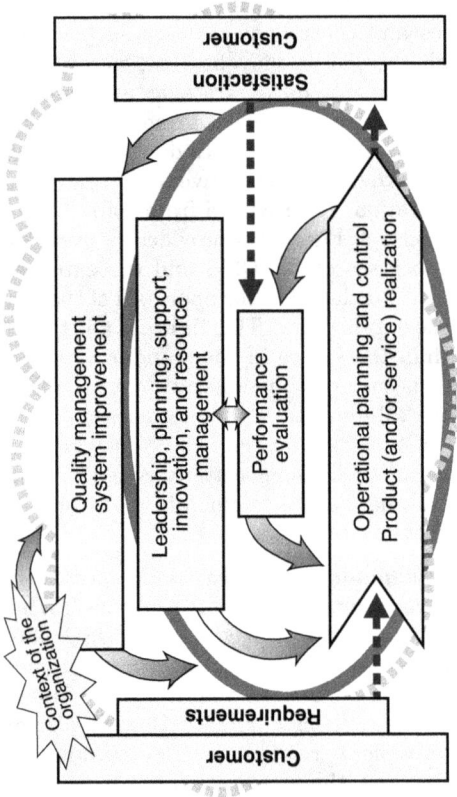

Figure 2.1   The processes of the QMS.

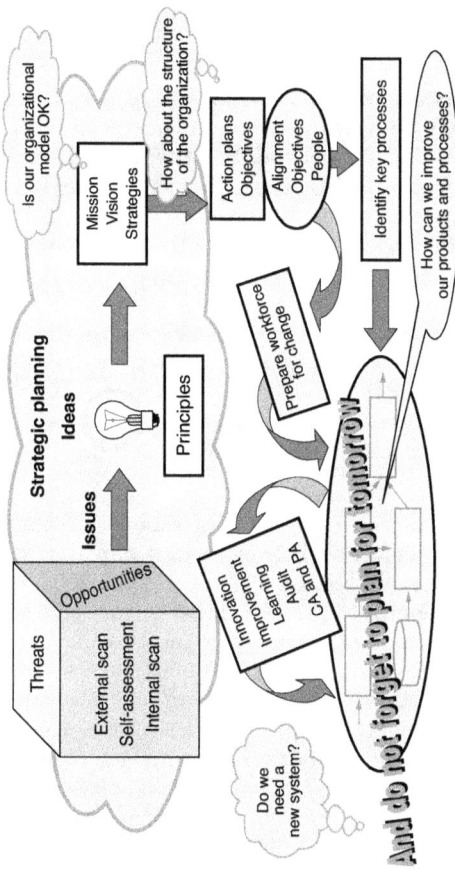

Figure 2.2 Context of the organization and consideration of threats.

the customer. This will be discussed further in Chapter 4.

• *Actions to address risks and opportunities.* Although we view the requirements around the concept of *risk* as vague, the intention of introducing risk into ISO 9001 is to encourage making preventive action an explicit element of the QMS. For many organizations, this will not be regarded as new. For others, new processes and thinking may be required. Although the wording in the standard may be vague and nonprescriptive to some, the concept is well worth attention if it has been ignored in the past. Clauses 4, 6, and 10 discuss risk and opportunities and the absence of preventive action requirements in detail.

• *Organizational knowledge.* Knowledge? A "new" concept? This requirement, although vague, could have been viewed as an implicit requirement of clause 6 in ISO 9001:2008 by some organizations. For others, the concept and the processes required to meet explicit and implicit requirements in ISO 9001:2015 will be new initiatives. It is interesting that the 2015 edition includes requirements related to knowledge management but does not include requirements related to *innovation*. The absence of innovation requirements was a conscious decision by the writers of the standard. Innovation is only mentioned in the introduction and in a note to one of the requirements in the improvements clause (clause 10). We bring your attention to the absence of requirements to consider innovation processes since we believe that innovation is and will continue to be an essential component of a QMS. Even though it is not an explicit requirement of ISO 9001:2015, organizations should consider addressing innovation requirements in their QMS from a sustainability viewpoint.

The comment above about innovation can be generalized. When planning the QMS for the organization, any and all requirements that may be essential to organizational sustainability that are not mandated by ISO 9001 should be considered. The organization needs to continually recognize that the objective is not to meet the requirements of a standard but rather to ensure customer satisfaction and to continually improve the performance of its QMS.

In addition to the concepts that may appear to be new in ISO 9001:2015, there have been changes in some of the "old" and "familiar" terminology that may be problematic for some organizations. Primary examples of "tweaked" terminology include:

- "Documented information" versus "documents and records"

- "Products and services" versus "products"

- Generic language

- "New" words or old words with definitions that have been modified

Following is an overview of these changes in terminology and a brief commentary on the rationale behind them:

- *"Documented information" versus "documents and records."* In ISO 9001:2015 the words "document(s)" and "record(s)" do not appear, having been replaced by "documented information." This change in terminology was dictated by the imposition of Annex SL on ISO 9001 with the thinking that "documented information" was a preferred way of referencing both documents and records. This terminology may be problematic in some organizations, especially those operating in regulated marketplaces. These words will be discussed in detail in the Chapter 4 content for clauses 4 thru 10, along with our recommendations for what to do or not do to address this new wording. By way of a sneak preview, we recommend doing absolutely nothing to change the way you refer to, think about, and use documents and records unless you have a reason to do so. We do not believe that the new wording used in the standard is a reason to change.

• *"Products and services" versus "products."*
In ISO 9001:2015 the word "products," which has
been in ISO 9001 since 1987, has been replaced
with the term "products and services," which
was dictated by Annex SL under the assumption
that "products" was not generic enough and not
as easy to understand for operations other than
manufacturing. Many disagree with this reason-
ing (including your authors), but we are stuck
with the new term.

• *Generic language.* Wherever possible
throughout the 2015 standard, there has been
an effort to use language that is as generic as
possible, under the assumption that, in the past,
ISO 9001 has been biased toward manufacturing
organizations.

• *"New" words or old words with tweaked
definitions.* We recommend that users of ISO
9001:2015 obtain a copy of ISO 9000:2015 *Quality management systems—Fundamentals and
vocabulary* to ensure a precise understanding of
the words (and how outside auditors will understand such words). Examples of some of the
words that may be new or have tweaked definitions whose formal ISO definitions should be
understood include:

  – Organization

- Interested party

- Risk

- Documented information

- Outsource

- Continual improvement

- Corrective action

- Correction

- Context of the organization

- Objective evidence

- Concession

ISO 9000:2015 is a normative reference in ISO 9001:2015, which means that the definitions in ISO 9000 shall be considered as requirements.

Other aspects of ISO 9001:2015 that may appear to be new include:

- Less prescriptive requirements

- Integrating the process approach and the systems approach to management

- Leadership emphasis

These aspects of ISO 9001:2015 are examples of either departure from the previous editions of ISO 9001 or areas of increased emphasis. Following are examples of how these aspects of the 2015 standard are integrated into the requirements:

• *Less prescriptive requirements*. An attempt was made in the 2015 edition to minimize prescriptive requirements. Where the 2008 edition contained requirements for a quality manual and a management representative, the 2015 edition has eliminated these requirements. Also, there are many places where documented information (that is, procedures) is not explicitly required. Remember that ISO 9001:2015 requires the organization to determine what documented information should be maintained, retained, or kept. So, the lack of a specific requirement in a particular clause does not mean the organization need not give careful consideration to determining what documented information (that is, procedures) is needed to ensure and demonstrate conformity. Users are also cautioned to consider auditing implications when deciding whether documented information (that is, a procedure or a record) should be required.

• *Integrating the process approach and the systems approach to management*. Although the process approach has been an explicit concept embedded in ISO 9001 since 2000, the 2015 standard clarifies and emphasizes the requirement to apply the process approach to the QMS.

• *Leadership emphasis.* Leadership receives much more explicit emphasis in ISO 9001:2015 than in past editions. Clause 5.1 states 10 specific items as requirements for management to demonstrate leadership and commitment to the QMS.

Chapter 3

Principles and Key Concepts

This chapter describes the eight quality management principles that were used as a basis for the development of ISO 9001:2015. It also discusses several other key concepts that are important and should be considered in the development and deployment of the QMS of an organization.

QUALITY MANAGEMENT PRINCIPLES

ISO 9001:2015, as was the case for the previous editions of ISO 9001, was developed with the seven quality management principles (QMPs) that are given in ISO 9000:2015 (clause 2.3) as a key input. While the QMPs help to form the foundation of ISO 9001, these principles are listed by their title only in clause 0.2 of ISO 9001:2015, which is not a part of the requirements. In ISO 9001:2015, the QMPs have been updated. Several principles were "tweaked" and *process approach* and *system approach to management* have been combined into one principle, which is now called *process approach*. The 2015 edition of the QMPs also recognizes that it is important to manage relationships with all interested parties, not only suppliers. This will be discussed further in Chapter 4.

The principles as they appear in ISO 9000:2015, along with the statements describing the principles, are as follows:

- *Customer focus.* The primary focus of quality management is to meet customer requirements and to strive to exceed customer expectations.

- *Leadership.* Leaders at all levels establish unity of purpose and direction and create conditions in which people are engaged in achieving the organization's quality objectives.

- *Engagement of people.* Competent, empowered, and engaged people at all levels throughout the organization are essential to enhance the organization's capability to create and deliver value.

- *Process approach.* Consistent and predictable results are achieved more effectively and efficiently when activities are understood and managed as interrelated processes that function as a coherent system.

- *Improvement.* Successful organizations have an ongoing focus on improvement.

- *Evidence-based decision making.* Decisions based on the analysis and evaluation of data and information are more likely to produce desired results.

- *Relationship management.* For sustained success, organizations manage their relationships with relevant interested parties, such as providers.

In addition to stating the principles, ISO 9000:2015 also includes a rationale of why each principle is important, a few examples of benefits associated with the principle, and examples of typical actions that an organization can take to improve performance when applying the principle.

Since the QMPs are a foundational element of the QMS of an organization, it is our strong recommendation that they be understood and embraced by everyone in the organization along with the vision and mission statements of the organization.

For more details on the quality management principles, run a search at http://www.iso.org.

CUSTOMERS—CUSTOMER FOCUS AND MEASURING SATISFACTION

The purpose of ISO 9001 is to achieve customer satisfaction by meeting customer requirements. While meeting requirements and preventing nonconformities have been fundamental to ISO 9001 since its initial issue in 1987, ISO 9001:2015 continues the enhanced emphasis on customers. There are key activities an organization needs to understand and implement that are related to customer focus and customer satisfaction in several clauses in the standard.

Examples include processes to ensure that:

- Top management demonstrates leadership and commitment with respect to customer focus by ensuring that:

  - Customer and applicable statutory and regulatory requirements are determined, understood, and consistently met.

  - The risks and opportunities that can affect conformity of products and services and the ability to enhance customer satisfaction are determined and addressed.

  - The focus on enhancing customer satisfaction is maintained.

- The quality policy contains a commitment to meet requirements. This includes meeting customer requirements.

- Customer feedback is obtained relating to products and services, including customer complaints.

- Information relative to customer satisfaction and feedback from relevant interested parties is obtained and provided as input to management review.

- Resources are provided to enhance customer satisfaction by meeting customer requirements.

- Customer requirements are determined and reviewed.

- Processes are required (clause 7.2.3) for communications with customers.

- Customers' perceptions of the degree to which their needs and expectations have been fulfilled are monitored.

Understanding this focus on the customer is critical to implementation of an effective ISO 9001 QMS. The organization should carefully think about the interrelated processes that are needed to meet the ISO 9001 requirements in a way that will enhance customer satisfaction.

PROCESS APPROACH— ACTIVITIES, PROCESS MANAGEMENT, AND THE SYSTEMS APPROACH TO MANAGEMENT

Of particular importance among the quality management principles is the *process approach*. People in any organization perform activities.

These activities are interrelated. The process approach involves managing the interrelated activities and associated resources together to achieve a particular output.

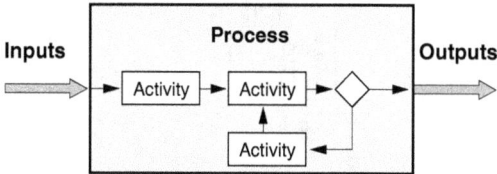

The process approach is basic to ISO 9001:2015. It encourages organizations to link interrelated value-adding processes. This linked system of processes results in the outputs that go to customers. Thus, the QMS needs to be composed of interrelated processes. Clause 0.3 of ISO 9001:2015 describes this concept of a system of processes within an organization.

This approach is easy to implement and has many advantages:

- It maintains focus on the creation of value by managing across the functional departments of the organization, thereby reducing the number and severity of quality problems that occur at department boundaries.

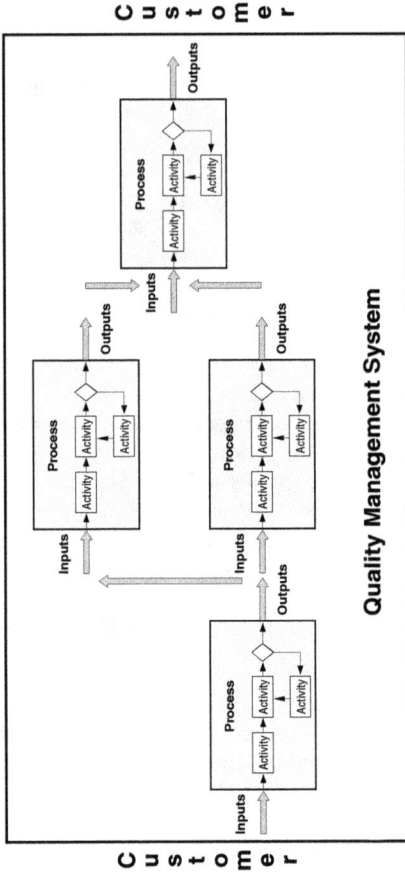

Quality Management System

- It helps the organization focus on what is important to both itself and its customers through measurement of product characteristics and performance of processes.

- It encourages open communications—based on facts supported with data—between internal customers, between internal suppliers, and between levels in the organization.

- It encourages continual improvement since any gaps between customer requirements and process performance are highlighted—quantitatively—and can be targeted for improvement efforts.

- It directly supports the principles of "leadership" and "involvement of people" with improvements involving everyone and every level of the organization.

- It provides a framework for managing innovation and accommodating changes to the QMS.

In summary, the process approach is very generic and applicable to all sectors and sizes of organizations. It helps create value by managing horizontally across functional departments, thereby reducing quality problems that typically occur between departments. And by tying key indicators of process performance to customer needs and supplier performance, it focuses on what is important to customers. It also strongly encourages continual improvement since it helps identify gaps between customer requirements and process performance. Finally, it involves everyone and every level of the organization in meeting requirements, customer satisfaction, and continual improvement.

The ISO committee responsible for the ISO 9000 family of standards has developed guidance on the process approach. For additional information on the process approach, see http://www.iso.org/tc176/sc02/public, section 2.

ALIGNMENT—QUALITY OBJECTIVES, PROCESS MEASUREMENT, AND COMMUNICATIONS

ISO 9001:2015 requires that quality objectives be measurable and aligned with the quality policy. It also requires that the measurable objectives be communicated, monitored, deployed, and updated (see clause 6.2).

The intent of this requirement is to ensure that responsibility and authority for key dimensions of the QMS are understood and deployed throughout the organization with the involvement of top management. Ensuring that objectives are measurable is intended to enhance improvement.

Clause 6.1 requires that the QMS be planned so that it meets the quality objectives. This means that the processes of the organization need to be operated, monitored, and measured with the organization's objectives in mind. Clauses 8 and 9 in several places require that processes be measured where applicable. It makes sense to measure the processes that are key to achieving the organization's quality objectives.

Objectives should be related to monitoring and measurement of the processes of the organization. Developing and implementing a process to address the establishment and deployment across the organization of measurable objectives that are aligned with the quality policy and truly drive the effectiveness of the QMS is an ongoing and challenging task, but it can be one of the major keys to success.

Properly deployed objectives with aligned process measures are important. Communication is also needed to ensure that the organization's people understand requirements (clause 7.4) and receive input on the effectiveness of the system (clauses 6, 7, 9, and 10).

CONTINUAL IMPROVEMENT

ISO 9001:2015 has a clear requirement for continual improvement of the QMS and its effectiveness. Improvement requirements are contained in several clauses of the standard, including clauses 6, 7, 9, and especially clause 10, which is titled "Improvement." Sustainable improvement of system effectiveness is accomplished by improving the processes of the system.

Examples of the requirements that contribute to improvement of the QMS include:

- Promulgating a quality policy with a commitment to continual improvement of the effectiveness of the quality management system. (Clause 5.2)

- Establishing and deploying measurable objectives at the relevant levels and functions of the organization. The objectives must be set with the commitment to improvement in mind. (Clauses 5 and 6)

- Collecting data. (Clause 9.1.3)

- Analyzing data. (Clause 9)

- Conducting meaningful management reviews to track progress, identify improvement opportunities, establish

priorities, and provide resources.
(Clause 9.3)

- Taking corrective action by eliminating the causes of nonconformities. (Clause 10.2)

- Assessing risks. (Clauses 4 and 6)

PROCESS CONTROL TO FACILITATE LEARNING, INNOVATION, AND IMPROVEMENT

Many of the requirements of ISO 9001 involve the concept that work is to be performed under controlled conditions. This concept is found throughout clause 8 and is best described in clause 8.5.1. Control of processes implies low variability, consistency of process performance, and a high level of product and process conformity to customer and internal requirements.

Designing process controls is critical to success of the QMS. If done wrong, the controls can stifle positive changes and continual improvement. But if the controls are properly integrated into the organization's processes, they can become important facilitators of learning, innovation, and improvement. It is not a matter of "balancing" control with learning, innovation, and change. Rather, it is a matter of developing controls so that they will promote learning, innovation, and improvement. When a control is being designed and implemented, it can be helpful to ask the question of how the control will facilitate learning, innovation, and continual improvement.

TOP MANAGEMENT AND PEOPLE INVOLVEMENT AT ALL LEVELS

Two of the most important of the quality management principles are *leadership* and *engagement of people*.

Top Management—Responsibility and Involvement

ISO 9001:2015 emphasizes the role of top management involvement in providing the leadership required for the development, deployment, and improvement of the QMS.

Clauses 5.1, 5.2, and 5.3 contain requirements for top management regarding their leadership and commitment to:

- The QMS

- Maintaining customer focus

- Establishing and maintaining the quality policy

- Ensuring that the responsibilities and authorities for relevant roles are assigned, communicated, and understood

The requirements for leadership and commitment in sub-clause 5.1.1 are explicit and clear. They are also extensive. Top managers need to be in control of the QMS through direct actions, where appropriate, or through delegation. The overall leadership of the QMS is not delegable but, in organizations of any real size, many components must be addressed by appropriate delegation. Care should be exercised to ensure that the assignments of responsibilities and authorities are clear and unambiguous. If two or more people share a responsibility or accountability, their individual roles need to be understood.

Top managers have a big role to play in ISO 9001:2015.

All People in the Organization

Everyone has a role in the deployment of a QMS that complies with the requirements of ISO 9001:2015. Some people are involved every hour of every day. Others may have periodic involvement. For a few people, involvement may be indirect or infrequent. The role of managers and supervisors is to find ways to get and keep everyone in the organization involved in the efficient and effective implementation and improvement of the QMS.

VISION AND MISSION

In addition to the above activities, to ensure that the QMS has a sound foundation, the organization needs to ensure that its vision and mission statements are current and understood throughout the organization. *Mission* is nothing short of the description of why the organization exists, and *vision* describes a picture of where the organization is going to be in the future. These are not trivial, soft topics. Instead, they are the essence of understanding an organization.

Maintaining a constant, clear understanding of the organization's mission and vision is an important prerequisite to sustainability because full alignment of the system and its processes with organizational needs requires an understanding of basic direction.

Mission: What is our business?

Vision: What do we want our business to be like in the future?

Chapter 4

What Are the ISO 9001:2015 Requirements and How Do We Comply?

This chapter provides a simplified explanation of the contents of ISO 9001:2015 and an overview of what is contained in the nonnormative clauses 0, 1, 2, and 3 of the standard. These clauses contain no requirements, which is why they are referred to as *nonnormative*. The titles of these clauses are (0) Introduction, (1) Scope, (2) Normative References, and (3) Terms and definitions.

The majority of this chapter addresses the requirements of clauses 4 through 10 of ISO 9001:2015. For each clause we provide the following:

What is the requirement? Provides a brief description of each requirement of 9001:2015.

Why do it? Gives a brief description from an organizational and management perspective of why the requirement should be addressed.

Implementation tips. Provides tips for your consideration as you develop processes to meet the requirements.

Questions to ask to assess conformity. Questions to consider asking that should be answered during implementation of processes to comply with requirements and when auditing the processes.

INTRODUCTION

The content of the introduction to ISO 9001:2015 exists to provide context, general understanding, and background. It discusses the intent of ISO 9001:2015 and the potential benefits to an organization of implementing a quality management system.

It also indicates that ISO 9001 is not intended to require or imply a requirement for uniformity of structure of a QMS, a need to align the documented information of an organization with the clause structure of the standard, or the need to conform with the terminology of the standard. In other words, each organization can be completely flexible to structure and implement and document its QMS any way it desires. Organizations should keep this front-end content of the standard in mind as they approach QMS deployment. The processes of an organization should be designed and implemented to add value and not to conform to an arbitrary structure of a standard.

The introduction also addresses some of the processes we highlighted in Chapter 3 that are of fundamental importance in the establishment of a sound foundation for the QMS.

CLAUSE 1 SCOPE

The scope clause of ISO 9001:2015 states in general terms what we presented in Chapter 1 as reasons for implementing processes that conform to ISO 9001:2015 requirements. The scope statement summarizes why organizations should use it, as follows:

- To demonstrate its ability to consistently provide products or services that meet customer and applicable statutory and regulatory requirements, and the organization's own internal requirements

- To enhance customer satisfaction through the effective application of the system, including processes for improvement of the

> system and the assurance of conformity
> to customer and applicable statutory and
> regulatory requirements

The scope also reminds users that ISO 9001:2015 is intended to be generic and applicable to all organizations, regardless of type, size, or products and services provided.

CLAUSE 2 NORMATIVE REFERENCES

The only normative reference included is ISO 9000:2015 *Quality management systems— Fundamentals and vocabulary*. This means, as we previously indicated, that the terms and definitions in ISO 9000:2015 shall be applied and used when developing, creating documented information, and deploying processes related to ISO 9001:2015.

We strongly encourage all users of ISO 9001 to obtain a copy of ISO 9000:2015.

CLAUSE 3 TERMS AND DEFINITIONS

The standard simply states that the terms and definitions given in ISO 9000:2015 apply.

CLAUSE 4 CONTEXT OF THE ORGANIZATION

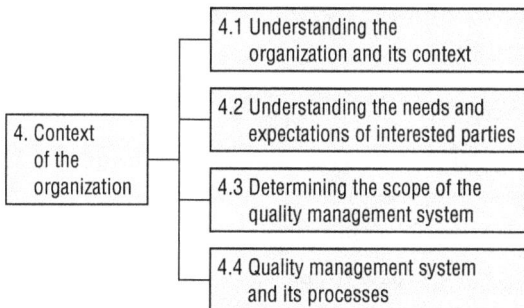

4. Context of the organization	4.1 Understanding the organization and its context
	4.2 Understanding the needs and expectations of interested parties
	4.3 Determining the scope of the quality management system
	4.4 Quality management system and its processes

CLAUSE 4.1 UNDERSTANDING THE ORGANIZATION AND ITS CONTEXT

What Is the Requirement?

This clause requires the organization to be thinking at both the strategic and tactical levels. This implies continuous thought about external and internal issues that are relevant to its purpose and its strategic direction. Neither "big picture" thinking nor detailed analysis is sufficient by itself. This clause mandates consideration of the internal and external issues that affect the ability of the organization to achieve its intended results when it develops and deploys its QMS.

Why Do It?

- Discover and understand threats and opportunities

- Consider threats

- Anticipate technology shifts

- Explore interactions between or among processes that may be related to external and internal issues

- Integrate the QMS into the strategic planning of an organization

Implementation Tips

- Have a process or processes for deciding what internal and external issues to consider and why

- Consider implementation of a self-assessment process

- List potential internal and external issues to consider

Questions to Ask to Assess Conformity

- Is there evidence that the organization has determined relevant external and internal issues?

- Is there evidence that the organization has monitored and reviewed information about relevant external and internal issues?

10 CLAUSE 4.2 UNDERSTANDING THE NEEDS AND EXPECTATIONS OF INTERESTED PARTIES

What Is the Requirement?

This clause introduces a requirement for the organization to determine parties that have an interest in and are relevant to its products and services and to determine the requirements of these parities. It further requires the organization to monitor and review information about these interested parties and their relevant requirements.

Why Do It?

- Attention to requirements of interested parties is a form of risk avoidance.

- "Interested parties" is new to ISO 9001, so processes related to understanding their relevant requirements require consideration.

- Inclusion of parties beyond the direct customer can impact the content of the QMS of the organization.

- Ideas for new products or services that can be offered to customers can be a desirable outcome from consideration of the needs or relevant requirements of interested parties.

Implementation Tips

- Define and deploy processes to determine whether interested parties are impacted by the activities of the organization.

- Document the processes for identification of interested parties and how relevance is assessed.

- Keep records (that is, documented information) of identification and relevance decisions.

- If there are interested parties identified, require consideration of actions to ameliorate or eliminate the extent of any impact.

- See Figure 4.1 for examples of organizations that could potentially be considered interested parties by the organization.

Questions to Ask to Assess Conformity

- What is the process for determining whether interested parties are impacted by the activities of the organization?

- What is the process for determining the requirements of interested parties?

- How is the monitoring and reviewing of the information about interested parties and their relevant requirements conducted?

10 CLAUSE 4.3 DETERMINING THE SCOPE OF THE QUALITY MANAGEMENT SYSTEM

What Is the Requirement?

This clause provides the requirements for determining the scope of the organization's QMS. It includes the following areas to consider when determining the QMS scope:

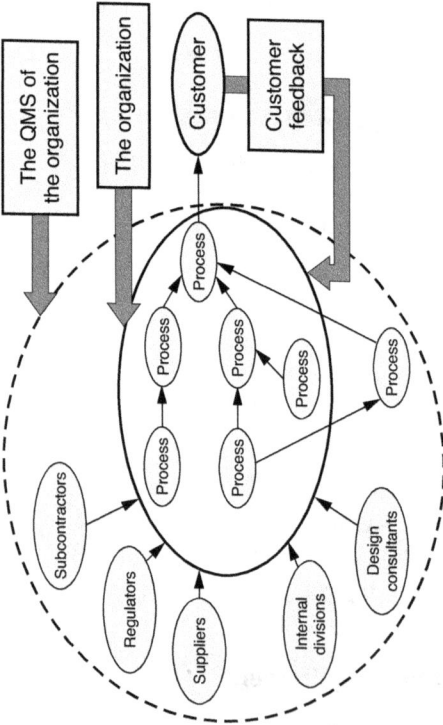

Figure 4.1   Examples of organizations that could potentially be considered interested parties.

- The external and internal issues relevant to its purpose and strategic direction

- The requirements of relevant interested parties

- The organization's products and services

- Any products and services affected by a claimed exclusion

- Maintenance of the scope as documented information

The documented information of the scope is required to state:

- How it is maintained

- Its availability

- The products and services embraced

- Any requirements that cannot be applied and therefore are excluded

Why Do It?

- To address the boundaries and applicability of the QMS and to ensure that the QMS scope is complete

- To ensure that the QMS will be adequate to achieve the objectives of the organization

- To require the organization to take a comprehensive view of its QMS and minimize or eliminate any exclusions

Implementation Tips

- Establish a process for scope development and maintenance

- Ensure comprehensiveness of scope while striving for brevity

- Retain documented information of the scope development process (that is, a procedure) and of the outputs of the process

- Define where and how the scope statement will be available and how it will be maintained

Questions to Ask to Assess Conformity

- How are the boundary conditions of the QMS established?

- How are external and internal issues considered?

- How are the requirements of relevant interested parties considered?

- What is the process and the output of the process for considering the products and services to be included in the scope of the organization?

- Does the scope statement state the products and services to which the QMS applies?

CLAUSE 4.4 QUALITY MANAGEMENT SYSTEM AND ITS PROCESSES

What Is the Requirement?

This is a very important clause in ISO 9001 that emphasizes the activities and the importance of the process approach. It has two untitled sub-clauses. It requires the organization to establish, implement, maintain, and continually improve a quality management system, including the processes needed and their interactions.

In the process of doing this, the organization is required to determine the processes needed for the QMS and their application throughout the organization. It needs to determine:

- The inputs required and outputs expected from processes

- The sequence and interaction of processes

- The criteria, methods, measurements, and performance indicators needed to ensure effective operation and control of the processes

- The resources needed

It shall also:

- Assign responsibilities and authorities for the processes

- Address and take action on the risks and opportunities that have been identified

- Implement any changes needed to ensure that the processes achieve their intended results

- Control changes to processes

- Improve the processes and the quality management system

To the extent necessary, the organization also shall maintain documented information to support the operation of its processes and retain documented information to have evidence that the processes are carried out under controlled conditions.

Why Do It?

- Process definition, management, control, and improvement is a fundamental ISO 9001 requirement.

- Process management and control is required to consistently meet customer requirements.

- Process management and control is required to have an effective prevention-based QMS.

- Expand past thinking on QMS content to include consideration of interested parties.

- Expand past thinking on QMS content to include consideration of the context of the organization

Implementation Tips

- Before contemplating the QMS and its processes, the organization should consider the foundations of the QMS.

- Review and update the mission and vision of the organization.

- Review the quality policy, the quality objectives, and the alignment of the policy and objectives with the overall vision and mission of the organization.

- Review overall consistency and alignment of vision, mission, strategic and tactical plans, policy, and objectives.

- Identify existing processes and determine what other processes, if any, must be developed.

- Ensure that there exists an understanding of customer needs and requirements.

- Map or flowchart the key processes, as appropriate.

- Ensure understanding of the interfaces and interactions of processes and potential areas for extra attention.

- For all processes, ensure consideration of process inputs, outputs, performance indicators, risk assessment, interactions, improvement, and controls.

Questions to Ask to Assess Conformity

- Have the processes needed for quality management been identified?

- Have the sequence and interaction of these processes been determined?

- Have criteria and control methods been determined for control of the processes in the quality management system?

- Is documented information available to support the operation and monitoring or measuring of the processes?

- Are processes measured, monitored, and analyzed, with appropriate actions taken to achieve planned results and improvement?

- Is the QMS established, documented, implemented, maintained, and continually improved?

- Has provision been made to ensure control of quality management system processes that are performed outside the organization?

CLAUSE 5 LEADERSHIP

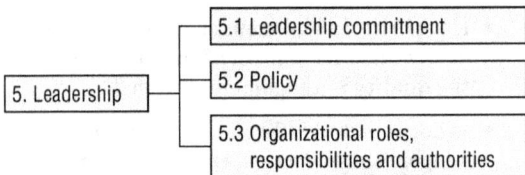

CLAUSE 5.1 LEADERSHIP AND COMMITMENT

What Is the Requirement?

Clause 5.1 has two sub-clauses:

    5.1.1 General

    5.1.2 Customer focus

The organization's top managers are required to demonstrate their leadership and commitment to the QMS by carrying out specific activities including at least the following:

- Taking accountability for QMS effectiveness

- Ensuring:

    - Establishment and maintenance of the quality policy

- The policy is compatible with the organization's context and strategic direction

- The policy is communicated, understood, and applied

- Integration of the QMS into business processes

- Availability of necessary resources for the QMS

- The QMS achieves intended results

- Customer requirements are determined and met

- Applicable statutory and regulatory requirements are determined and met

- Risks and opportunities that may affect conformity of products and services are determined and addressed

- Risks and opportunities related to the ability to enhance customer satisfaction are determined and addressed

- There is a focus on consistent delivery of products and services that meet customer requirements

- Provision of direction and support to enhance QMS effectiveness

- — Promotion of continual improvement

- — Understanding of the process approach in the organization

- — Communication of the importance of effective quality management and conformity to the QMS requirements

- — Support and encouragement to other managers to demonstrate leadership in their area of responsibility

- Participation in management review (addressed in the discussion on Clause 9)

- Top management also has requirements to demonstrate leadership and commitment to customer focus by ensuring that:

  - — Customer and any applicable regulatory requirements, if any, are determined and met.

  - — Risks and opportunities identified that can impact conformity to requirements or customer satisfaction are determined and appropriate actions are taken.

  - — A continual focus is maintained on customer satisfaction.

Why Do It?

- Provide top management control of and interaction with the QMS

- Ensure that responsibilities, accountabilities, and authorities are clear and unambiguous

- Enhance communication with and involvement of all staff throughout the organization

- Enhance risk assessment and exploration of opportunities

- Emphasize customer satisfaction

- Integrate quality management into the tactical activities of the organization

- Ensure alignment of process implementation and objectives

Implementation Tips

The question of how to improve top management leadership and involvement in quality management has plagued (and perplexed) quality professionals for a long time and has been a source of misplaced irritation. We say misplaced irritation because, rather than QA folks complaining about the lack of top management involvement, we should be looking inward at our own failure to establish effective communications with top management.

What should we be doing? Of course, all we can do here is provide a starting point for your own introspection to find processes that will work in your organizations. One area to consider is how messages and requests for top management involvement are presented or communicated. Is interaction regarding issues related to quality management posited in the language of management? Do we talk about error rates or costs? Do we discuss acceptable quality levels (AQLs) or earnings per share (EPS)? Have we made quality costs visible in monetary terms? Do we understand how to integrate the impacts of product, service, and process quality into a balanced scorecard? Are we knowledgeable about the impact of the London Interbank Offered Rate (LIBOR) or the Federal Reserve Board (FRB) on our organization? We need to embed in our minds that the language of top managers is *money*.

Quality professionals should consider development of processes that communicate requirements and needs in the language of management. By way of analogy, if an individual is in a bistro in a small village in southern France and is hungry, a successful attempt to obtain food would have a much higher probability of success if communication were conducted in either the French language or sign language than attempting to communicate in English.

The current edition of ISO 9001 does provide a much more explicit listing of top management requirements related to how they can provide leadership and commitment to the QMS, but effectiveness will not occur until and unless the benefits are evident. *It is a primary role of the quality professionals to make such benefits evident.*

Top management engagement in enhancing customer focus could be "packaged" in terms of enhancing market share and lowering various components of costs. This would position the importance of maintenance of focus on customers as having a significant financial component.

It is also meaningful to review the *quality management principles* (QMPs), listed in clause 0.2 of ISO 9001:2015, which are one element of the foundation of ISO 9001. Leadership is presented in the quality management principle on leadership—QMP 2—which is included in more detail in clause 2.3.2 of ISO 9000:2015. Clause 2.3 of ISO 9000:2015 includes a statement of all the QMPs along with their rationale, the key benefits associated with each QMP, and the possible actions organizations can consider related to each QMP.

Questions to Ask to Assess Conformity

- Is there evidence that top management is demonstrating leadership and commitment with respect to the quality management system as required by the 10 activities articulated in clause 5.1.1?

- Is there evidence that top management is demonstrating leadership and commitment to ensuring and maintaining customer focus?

CLAUSE 5.2 POLICY

What Is the Requirement?

Clause 5.2 has two sub-clauses:

5.2.1 Establishing the quality policy

5.2.2 Communicating the quality policy

Top managers shall establish the quality policy, review it, maintain it, and ensure that it:

- Is appropriate for the organization's purpose and context

- Provides a framework for setting and reviewing quality objectives

- Includes commitment to meet requirements

- Includes commitment to improvement

- Is available and maintained as documented information

- Is communicated, understood, and applied within the organization

- Is available to relevant interested parties

Why Do It?

- To ensure that the quality policy is established, communicated, and understood within the organization and that it is applied and available to all relevant interested parties

- To ensure that the quality policy is appropriate and provides a framework for setting and reviewing objectives

- To ensure that the quality policy is maintained

Implementation Tips

- Ensure that the quality policy is consistent with strategic and tactical plans

- Promote the use of quality policy in marketing initiatives to enhance sales of products and services, as appropriate

- Tie the quality policy to the overall mission and vision of your organization

- Make it easy for people to describe the policy in their own words without quoting it word for word

- Use words and content that reflect the culture of your organization

Questions to Ask to Assess Conformity

- Has a quality policy been developed?

- Has top management determined that the quality policy meets the needs of the organization and its customers?

- Does the quality policy include commitment to meeting requirements and commitment to continual improvement?

- Does the quality policy provide a framework for establishing and reviewing the quality objectives?

- Is the policy communicated to and understood by all in the organization?

- Are the members of the organization clear as to their roles in carrying out the policy?

- Is the quality policy included in the documented information process?

- Is the quality policy reviewed for continuing suitability?

4 CLAUSE 5.3 ORGANIZATIONAL ROLES, RESPONSIBILITIES AND AUTHORITIES

What Is the Requirement?

An important role for top management is to ensure that responsibilities and authorities are assigned, communicated, and understood for:

- Designing, deploying, maintaining, and improving the QMS content in compliance with internal and ISO 9001 requirements

- Monitoring processes to ensure they deliver intended outputs

- Promoting customer focus everywhere in the organization

- Maintenance of the QMS when changes are planned and implemented

- Reporting on the performance of the QMS

- Reporting on opportunities for improvement

- Reporting on the need for changes to the QMS

Why Do It?

- To indicate that clarity is important for all key personnel

- To ensure understanding of who is responsible for identifying, analyzing, and approving nonconformities

- To have a clear understanding of process handoffs

Implementation Tips

- Create and maintain organization charts

- Use job descriptions

- Use organization charts to show functional relationships

- Use Gantt charts to show responsibilities for projects

- Include in documented procedures who (by job function) is responsible for activities

Questions to Ask to Assess Conformity

- Are the organization's functions and interactions defined and communicated to facilitate effective quality management?

- Are responsibilities and authorities defined and communicated to facilitate effective quality management?

- Is there a process to ensure that employees understand the importance of meeting customer, regulatory, and statutory requirements?

- Is there evidence that top management ensures that the responsibilities and authorities for relevant roles are assigned, communicated, and understood within the organization?

CLAUSE 6 PLANNING

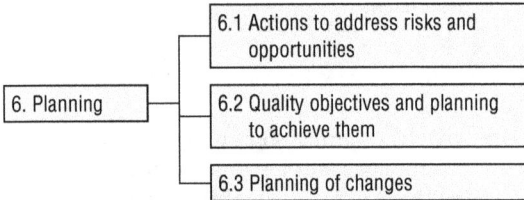

```
                    ┌─────────────────────────────────┐
                    │ 6.1 Actions to address risks and│
                    │     opportunities               │
                    └─────────────────────────────────┘
┌──────────────┐    ┌─────────────────────────────────┐
│ 6. Planning  │────│ 6.2 Quality objectives and      │
└──────────────┘    │     planning to achieve them    │
                    └─────────────────────────────────┘
                    ┌─────────────────────────────────┐
                    │ 6.3 Planning of changes         │
                    └─────────────────────────────────┘
```

CLAUSE 6.1 ACTIONS TO ADDRESS RISKS AND OPPORTUNITIES

(Including untitled sub-clauses 6.1.1 and 6.1.2)

What Is the Requirement?

Clause 6.1 is interrelated with clauses 4.1 (Understanding the organization and its context) and 4.2 (Understanding the needs and expectations of interested parties). It requires consideration of issues relevant to the purposes of the organization (see clause 4.1) and determination of the risks and opportunities that need to be addressed in planning the QMS to enable the organization to achieve intended results, prevent or reduce undesired effects, and achieve improvement.

The planning activities are required to address how to integrate the actions into the QMS and how to evaluate effectiveness.

Clause 6.1 states that actions taken to address risks and opportunities shall be proportionate to the potential impact on conformity of products and services.

There is also a Note 1 to clause 6.1.2 that offers suggestions for addressing risks and opportunities. Note 2 to clause 6.1.2 indicates that opportunities can lead to the adoption of new practices, launching new products, opening new markets, addressing new clients, building partnerships, using new technology, and other desirable and viable possibilities to address the organization's or its customers' needs. Remember that Notes do not contain requirements. They are intended to provide information to enhance understanding.

Why Do It?

- Compliance with the requirements of clause 6.1.1 requires the organization to consider risks and opportunities as well as other interested parties when planning the QMS, thereby providing potential areas for expanding its breadth and depth.

- Considering threats and interested parties when planning the QMS can be important from a sustainability viewpoint.

- Consideration of the actions the organization plans to take to address risks and opportunities can be a powerful tool in avoiding experiencing the risks.

- Relevant interested parties beyond customers can have a negative impact on an organization if their needs and expectations are not considered.

Implementation Tips

- Dealing in a systematic way with known uncertainties of the organization will improve the opportunity for the organization to avoid situations that could be problematical, and also may present great opportunities.

- Achieving an understanding of external and internal issues and establishing a process to recognize changes in this information over time can be a useful input into the strategic planning process.

- Some areas worthy of consideration when contemplating potential internal issues include:

  – Internal audit results and self-assessment results

- Analysis of quality cost data

- Analysis of technology trend information

- Competitive analysis

- Results of customer reviews, audits, and feedback

- Actual versus intended internal values and culture

- Organizational performance

- Best practices of the organization and comparisons with industry benchmarks

- For identification of external issues, organizations can consider analysis of:

  - The economic environment

  - International trade conditions, processes, and barriers.

  - Competitive products and services

  - Technology trends

  - Regulatory changes

  - Opportunities and conditions related to a wide variety of outsourcing strategies

- Benchmarking of processes with organizations using similar processes can also be effective.

- Consider developing and deploying a self-assessment process to identify important and relevant internal and external issues.

Questions to Ask to Assess Conformity

- Have the risks and opportunities that need to be addressed been identified?

- Have undesired effects been considered?

- Has improvement related to both risks and opportunities and interested parties been considered?

- Have external and internal issues been considered?

- Is there evidence that information about external and internal issues is being monitored and reviewed?

- Have interested parties that are relevant to the quality management system been determined?

- Have the requirements of interested parties been determined?

- Is there evidence that information about interested parties and their relevant requirements is being monitored and reviewed?

🔑1 🔑3 CLAUSE 6.2 QUALITY OBJECTIVES AND PLANNING TO ACHIEVE THEM

(Including untitled sub-clauses 6.2.1 and 6.2.2)

What Is the Requirement?

Quality objectives shall be established at relevant functions and levels.

There is a laundry list of characteristics that shall be applicable to objectives. They shall be:

- Consistent with the quality policy

- Measurable

- Based on applicable requirements

- Relevant to conformity of products and services and to enhancement of customer satisfaction

- Monitored

- Communicated

- Updated, as appropriate

Retention of documented information on quality objectives is required.

Why Do It?

- To clearly align the quality objectives with the quality policy throughout the organization

- To align everyday work with the quality policy

- To provide targets against which results can be measured and performance can be reviewed

- To drive improvement; what gets measured, gets done

Implementation Tips

- Be sure your quality objectives align with the quality policy

- Consider brainstorming to develop initial ideas related to the objectives

- Avoid conflicting priorities by ensuring that quality objectives align with other objectives of the organization

- Align measures with the quality objectives

- Focus on measuring the right things

- Change the objectives, if required, as situations change and as experience is gained with the system

- Consider using a balanced scorecard, dashboard, or similar process for monitoring progress

Questions to Ask to Assess Conformity

- Have quality objectives been established at each relevant function and level in the organization?

- Do quality objectives include those needed to meet requirements for the organization's products or services?

- Are the quality objectives consistent across the organization?

- Are the quality objectives measurable?

- Are the quality objectives reviewed against the quality policy?

🔧 CLAUSE 6.3 PLANNING OF CHANGES

What Is the Requirement?

When there is a need for changes to the QMS, the changes shall be carried out under controlled conditions (that is, in a planned and systematic manner) that consider:

- What will be done and why
- Potential adverse consequences
- Maintenance of the integrity of the QMS
- What resources will be required
- Allocation or reallocation of responsibilities and authorities

Why Do It?

- Ensure the ongoing integrity of the QMS
- Maintain control of processes to ensure that requirements continue to be met
- To eliminate or at least minimize product and process variability

- To ensure that products and services continue to conform to requirements and meet internal and customer expectations

Implementation Tips

- When changes are required to any element of the QMS, consider having a process that requires exercising controls such as design review, verification, and validation of proposed changes.

- Ensure that processes that control changes to the QMS require changes to be carried out in a planned and systematic manner, that is, under controlled conditions.

Questions to Ask to Assess Conformity

- How does the organization ensure that changes to the quality management system are carried out in a planned and systematic manner?

- Are changes to the QMS discussed and reviewed in the management review meetings?

CLAUSE 7 SUPPORT

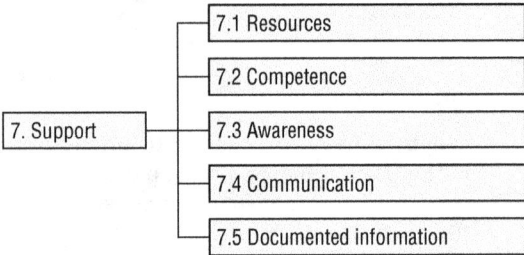

```
                      ┌─────────────────────────────┐
                    ┌─┤ 7.1 Resources               │
                    │ └─────────────────────────────┘
                    │ ┌─────────────────────────────┐
                    ├─┤ 7.2 Competence              │
                    │ └─────────────────────────────┘
  ┌────────────┐    │ ┌─────────────────────────────┐
  │ 7. Support ├────┼─┤ 7.3 Awareness               │
  └────────────┘    │ └─────────────────────────────┘
                    │ ┌─────────────────────────────┐
                    ├─┤ 7.4 Communication           │
                    │ └─────────────────────────────┘
                    │ ┌─────────────────────────────┐
                    └─┤ 7.5 Documented information   │
                      └─────────────────────────────┘
```

CLAUSE 7.1 RESOURCES

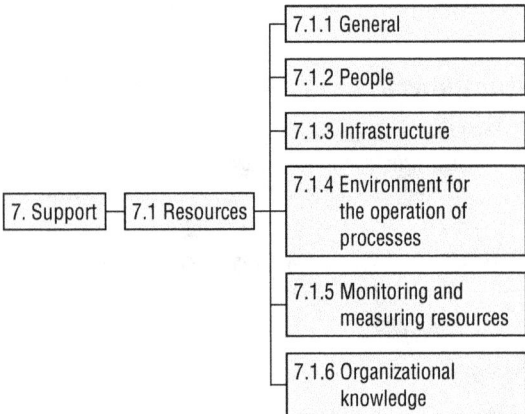

```
                                  ┌────────────────────────┐
                                ┌─┤ 7.1.1 General          │
                                │ └────────────────────────┘
                                │ ┌────────────────────────┐
                                ├─┤ 7.1.2 People           │
                                │ └────────────────────────┘
                                │ ┌────────────────────────┐
                                ├─┤ 7.1.3 Infrastructure   │
                                │ └────────────────────────┘
                                │ ┌────────────────────────┐
 ┌────────────┐ ┌─────────────┐ │ │ 7.1.4 Environment for  │
 │ 7. Support ├─┤7.1 Resources├─┼─┤   the operation of     │
 └────────────┘ └─────────────┘ │ │   processes            │
                                │ └────────────────────────┘
                                │ ┌────────────────────────┐
                                ├─┤ 7.1.5 Monitoring and   │
                                │ │   measuring resources  │
                                │ └────────────────────────┘
                                │ ┌────────────────────────┐
                                └─┤ 7.1.6 Organizational   │
                                  │   knowledge            │
                                  └────────────────────────┘
```

What Is the Requirement?

Clause 7.1 has six sub-clauses:

    7.1.1 General

    7.1.2 People

    7.1.3 Infrastructure

    7.1.4 Environment for the operation
       of processes

    7.1.5 Monitoring and measuring resources

    7.1.6 Organizational knowledge

Each of these sub-clauses addresses requirements for ensuring that the resources necessary in meeting both customer and internal requirements are considered and deployed as appropriate. The attention to ensuring the availability of resources is heightened in the 2015 edition of ISO 9001. Details of the contents of the sub-clauses are as follows:

- Clause 7.1.1 requires the organization to determine and provide the resources needed to establish, implement, maintain, and improve the quality management system. It also requires consideration of the capabilities of, and constraints on, existing internal resources and the needs of external providers.

- Clause 7.1.2 requires the organization to determine and provide the necessary persons for the effective operation and control of the QMS and its processes.

- Clause 7.1.3 requires the organization to provide and maintain the infrastructure necessary for the operation of its processes.

- Clause 7.1.4 simplifies the requirements related to the process environment. It simply requires the organization to provide and maintain the environment necessary for the operation of its processes. The Note to 7.1.4 is long and nonnormative (that is, it contains no requirements) and will not be very helpful to many users.

- Clause 7.1.5 (which includes sub-clauses 7.1.5.1 General and 7.1.5.2 Measurement traceability) incorporates the requirements for monitoring and measuring resources. The requirements can be summarized as follows:

  − Determine the resources needed for valid and reliable monitoring and measuring of processes.

  − Ensure that the resources provided are suitable.

- Ensure that the resources provided are maintained.

- Retain appropriate documented information as evidence of fitness for purpose of the monitoring and measurement resources.

- Where measurement traceability is a requirement, measuring instruments shall be verified or calibrated at specified intervals or prior to use against measurement standards traceable to international or national measurement standards.

- The basis used for calibration or verification shall be retained as documented information if no standard exists.

- Measuring instruments shall be identified.

- Measuring instruments shall be safeguarded from adjustments, damage, or deterioration.

- When an instrument is found to be defective, determine whether there is any adverse effect on the validity of previous measurement results.

Old-school metrologists will be either amused or annoyed by a few of the subtle changes that have been made, including replacing the word "equipment" with "resources."

It is interesting to us that confirmation of software, which is integral to much contemporary instrumentation, is not mentioned as a requirement. We believe that the intent of the words ". . . ensure the resources provided are suitable . . ." does require organizations to include determination that software that is an element of a measuring or monitoring system does function in accord with requirements (both doing what it is supposed to do and not doing what it is not intended to do).

- Clause 7.1.6, Organizational knowledge, is a new requirement under ISO 9001. The requirement, while somewhat vague and subject to a wide spectrum of understanding (and hence creating a potential for many issues related to auditability), can be summarized as follows:

  - Determine the knowledge necessary for the operation of processes

- Maintain necessary knowledge and make it available, as appropriate

- Consider the current knowledge inventory of the organization and determine how to acquire or access the additional knowledge necessary to address changing needs and trends

Additional information related to this requirement can be found in Annex A (A.7 Organizational knowledge).

Why Do It?

- To ensure the availability of the human and capital resources needed to establish, implement, maintain, and improve the quality management system

- To ensure that the work environment is suitable

- To ensure that adequate infrastructure is determined, provided, and maintained

- To ensure that currently required organizational knowledge is obtained, and to plan to acquire any necessary additional knowledge and required updates

- To ensure QMS effectiveness

- To ensure that the needed resources are provided to obtain valid and reliable results when monitoring or measuring is used to verify the conformity of products and services to requirements

- To ensure that resources are properly applied for the improvement of the overall organization and its stakeholders

Implementation Tips

- Identify the processes of the QMS

- Consider resource requirements for processes (personnel, time, buildings, equipment, utilities, materials, supplies, instruments, software, transport facilities, other infrastructure)

- Consider those parts of the organization that impact product quality, not just the resources needed to operate the "quality department"

- Consider short- and long-term resource needs; timing may be critical

- Integrate longer-term needs into the organization's strategic and capital plans

Questions to Ask to Assess Conformity

- How are required resources determined?

- What needs to be obtained from external providers?

- How are infrastructure requirements determined and assessed for adequacy?

- How are environmental requirements necessary for the operation of processes and to achieve conformity of products and services determined and assessed for adequacy?

- Has the organization identified the measurements to be made?

- Has the organization retained appropriate documented information as evidence of fitness for purpose of monitoring and measurement resources?

- Are monitoring and measurement devices calibrated and adjusted periodically or before use against devices traceable to international or national standards?

- Is the basis used for calibration recorded when traceability to international or national standards cannot be done since no standards exist?

- Are monitoring and measurement devices protected from damage and deterioration during handling, maintenance, and storage?

- How is determination of knowledge necessary for the operation of processes obtained?

- If changing needs and trends are anticipated, does the organization consider how to acquire or access the necessary additional knowledge?

CLAUSE 7.2 COMPETENCE

What Is the Requirement?

Clause 7.2 requires the organization to:

- Determine the competence required for persons doing work that affects the performance and effectiveness of the QMS

- Ensure competence on the basis of appropriate education, training, or experience

- Take actions to acquire necessary competence, and evaluate effectiveness of actions taken

- Retain appropriate documented information as evidence of competence

Why Do It?

- To ensure that people have the capability to satisfy customers by providing product that meets customer requirements.

- To ensure QMS effectiveness.

- To ensure that people have the capability to make the QMS effective.

- To ensure that people have the capability to continually improve the effectiveness of the QMS by improving its processes.

- To ensure that competence requirements apply to all personnel since virtually everyone in the organization does work that affects quality performance.

Implementation Tips

- Use some form of written job requirements (that is, job descriptions).

- Competencies required should make sense for each job type; find the right combination of education, training, skill, and experience.

- Some job requirements may be only a performance standard (for example, to produce a quantity of conforming product per day).

- Establish processes to ensure that newly hired or transferred personnel have the competence to perform assigned tasks to requirements.

Questions to Ask to Assess Conformity

- How does the organization determine the competence necessary for person(s) doing work?

- How does the organization ensure that personnel are competent on the basis of appropriate education, training, and experience?

- Is appropriate documented information available as evidence of competence?

- Where action is taken to acquire necessary competence, is the effectiveness of the actions taken evaluated?

CLAUSE 7.3 AWARENESS

What Is the Requirement?

Clause 7.3 requires individuals doing work to have awareness of:

- The quality policy

- Relevant quality objectives

- The contribution they provide to the effectiveness of the system, including improved quality performance

- The implications of not conforming with requirements

Why Do It?

- To ensure that people understand why their work is important and how they contribute to the success of the organization

- To ensure awareness of the quality policy throughout the organization

- To ensure awareness of the consequences of not conforming to QMS requirements

Implementation Tips

- Create and deploy a process to ensure consistent conformity to requirements

- To create awareness of the quality policy and objectives and the importance of each employee, consider activities such as the following:

  – E-mails from top management to all employees

  – Top management interaction with all employees

  – Meetings hosted by top management to inform employees of the "state of the organization"

  – Postings on bulletin boards

  – Letters to all employees

- Retain documented information of actions taken

Questions to Ask to Assess Conformity

- Are all members of the organization aware of the quality policy and its meaning?

- Are all members of the organization aware of how their actions relate to achievement of the organization's objectives?

- How is the importance of the contribution of each individual in the organization to the effectiveness of the quality management system communicated?

CLAUSE 7.4 COMMUNICATION

What Is the Requirement?

This clause requires the organization to determine both the internal and external communications relevant to the QMS, including:

- What it will communicate

- When to communicate

- With whom to communicate

- How to communicate
- Who communicates

Why Do It?

- To ensure customer satisfaction
- To solve problems quickly and early
- To increase business with customers by "being in touch" on a regular basis
- To enhance alignment of activities throughout the organization

Implementation Tips

- Make the level and form of customer contact consistent with the products you sell and the volume of business with a customer
- Tie the processes needed to meet this requirement to your processes for addressing requirements relating to monitoring information on customer satisfaction (see clause 9.1.2)
- Consider processes for contracts or order handling, including changes, and any customer feedback regarding your products or services, including customer complaints

Questions to Ask to Assess Conformity

- How has the organization determined its internal and external communications channels?

- Is responsibility clear regarding who will communicate what and when and how?

- Are there effective processes in place to facilitate communication with customers about product information, inquiries, contracts, order handling (including amendments or changes), and customer feedback, including customer complaints?

- Have you linked the processes for communicating with customers with those for monitoring customer satisfaction?

✌ CLAUSE 7.5 DOCUMENTED INFORMATION

What Is the Requirement?

Clause 7.5 has three sub-clauses: 7.5.1, 7.5.2, and 7.5.3, with 7.5.3 having two untitled sub-sub-clauses, 7.5.3.1 and 7.5.3.2. The clauses are titled:

7.5.1 General

7.5.2 Creating and updating

7.5.3 Control of documented information

The most significant change in ISO 9001:2015 is in the nomenclature. The previous terms of "documents" and "records" have been replaced with "documented information."

Since this change in terminology can have a significant impact on the overall documentation of the QMS (that is, all its procedures and records) let us understand the ISO 9000:2015 definition of *documented information*: ". . . information required to be controlled and maintained by an organization."

The essence of the requirements for documented information is that the QMS shall include:

- All documented information required by ISO 9001:2015

- Documented information that the organization determines as necessary for effective operations

When documented information is created and updated, the organization shall ensure appropriate:

- Identification and description (for example, a title, date, author, or reference number)

- Format (for example, language, software version, graphics), and media (for example, paper, electronic)

- Review and approval for suitability and adequacy

Documented information required by the quality management system shall be controlled to ensure:

- Availability and suitability for use, where and when it is needed

- Adequate protection

For the control of documented information, the organization shall address the following activities, as applicable:

- Definition and maintenance of distribution, access, retrieval, and use

- Storage and preservation

- Control of changes

- Retention and disposition

- Identification and control of documents of external origin (such as industry and customer specifications and standards)

- Protection from unintended alteration of documented information retained as evidence of conformity

Why Do It?

- To ensure consistent performance of all activities affecting quality

- To ensure that controls are in place to issue and approve documents

- To ensure that changes in requirements are communicated to those who must implement them

- To prevent the use of obsolete information

- To make certain that people have up-to-date instructions and requirements

Implementation Tips

- Define the types of documents in your system.

- Determine whether any change in nomenclature is required or desired.

- Define the types of documentation that come from other organizations (for example, standards, customer documents).

- Maintain a clear differentiation between documents and records even though the standard does not require such a distinction.

- Where the term "maintain documented information" is used, the intent is that the documented information be kept up to date.

- Where the term "retain documented information" is used, the intent is similar to the old requirement for a record.

- Define the control process appropriate for each type of document; for example, the requirements for computer-based documentation may be different from those for paper drawings.

- Records are a special type of document and require different controls.

- See Table 4.1 for reference when considering the terminology for documented information.

- Users of the standard should also be aware that, although the primary requirements for documented information are in clause 7.5, documented information is also addressed in at least 21 other places in clauses 4 through 10. See Table 4.2 for examples of clauses that mention documented information.

Table 4.1 Terminology for documented information.

Old term	New term	Purpose of documented information	Changes and corrections
Documented procedure	Maintain documented information	Provide up-to-date information for process operations	Keep up to date with changes
Record	Retain documented information	Provide objective evidence of activities conducted	No changes, but corrections may be permitted

In Table 4.2:

- M indicates a requirement to maintain documented information related to a requirement. In earlier versions of ISO 9001 this would have been expressed as a requirement for a documented procedure(s).

- R indicates a requirement to retain documented information related to a requirement. In earlier versions of ISO 9001 this would have been expressed as a requirement for records.

Table 4.2 Examples of clauses referencing documented information.

Documented information	Documented information	Documented information
4.3 M	8.1 D & K	8.5.2 R
4.4.2 M & R	8.2.3.2 R	8.5.3 R
5.2.1 M	8.3.2 D	8.5.6 R
5.2.2 M	8.3.3 R	8.6 R
6.2.1 M	8.3.4 R	8.7.2 R
7.1.5.1 R	8.3.6 R	9.1.1 R
7.1.5.2 R	8.4.1 R	9.2.2 R
7.2 R	8.5.1 D & A	9.3.3 R
7.5.3.2 R	8.3.6 R	10.2.2 R

- K indicates requirements for both maintaining and retaining documented information.

- D indicates a requirement for the organization to determine the documented information needed to effectively conform to a requirement.

- A indicates that documented information shall be "available"

Questions to Ask to Assess Conformity

- Has the documented information required by this standard been established?

- Is all documented information determined by the organization as being necessary for the effectiveness of the quality management system available?

- Is documented information approved for adequacy prior to use?

- Is documented information reviewed and updated as necessary?

- Are changes to documented information reapproved to ensure adequacy prior to use?

- Is the current revision status of documented information maintained?

- Are relevant versions of applicable documented information available at points of use?

- Is there a process to ensure that documented information remains legible, identifiable, available when and where required, and retrievable?

- Is documented information of external origin identified?

- Is obsolete documented information that is retained for any purpose suitably identified to prevent unintended use?

- Is documented information protected (for example, from loss of confidentiality, improper use, or loss of integrity)?

CLAUSE 8 OPERATION

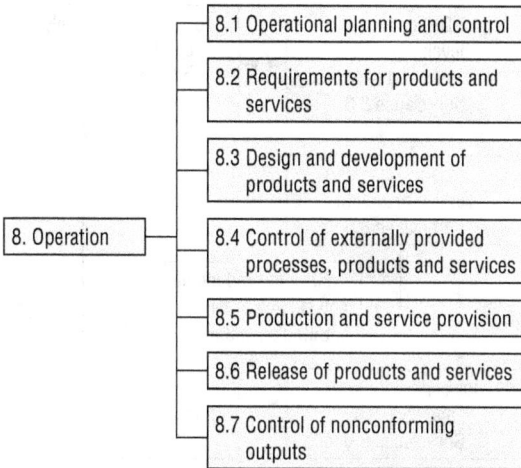

```
                    ┌──────────────────────────────┐
                ┌───┤ 8.1 Operational planning and control │
                │   └──────────────────────────────┘
                │   ┌──────────────────────────────┐
                ├───┤ 8.2 Requirements for products and │
                │   │     services                 │
                │   └──────────────────────────────┘
                │   ┌──────────────────────────────┐
                ├───┤ 8.3 Design and development of │
                │   │     products and services    │
                │   └──────────────────────────────┘
┌────────────┐  │   ┌──────────────────────────────┐
│ 8. Operation ├──┼───┤ 8.4 Control of externally provided │
└────────────┘  │   │     processes, products and services │
                │   └──────────────────────────────┘
                │   ┌──────────────────────────────┐
                ├───┤ 8.5 Production and service provision │
                │   └──────────────────────────────┘
                │   ┌──────────────────────────────┐
                ├───┤ 8.6 Release of products and services │
                │   └──────────────────────────────┘
                │   ┌──────────────────────────────┐
                └───┤ 8.7 Control of nonconforming │
                    │     outputs                  │
                    └──────────────────────────────┘
```

Figure 4.2 provides a picture in flow diagram format of the contents of clause 8 and the interrelationships of the seven sub-clauses.

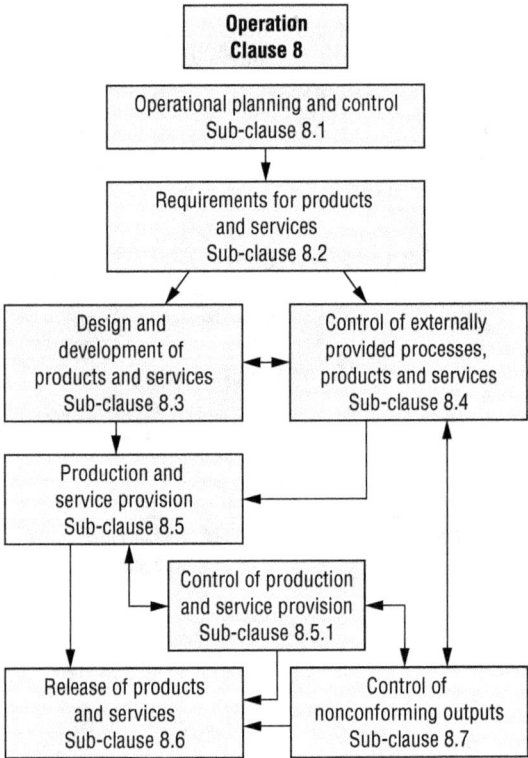

Figure 4.2 Clause 8 relationship diagram.

⚒ CLAUSE 8.1 OPERATIONAL PLANNING AND CONTROL

What Is the Requirement?

Clause 8.1 requires that operations be conducted through processes that are planned and controlled regardless of whether the organization or an outside party performs the processes. Requirements for products and services must be determined and criteria established for acceptance. Identification of resources needed to achieve conformity is required. Planned changes are required to be controlled and action taken to mitigate the effects of unintended consequences of changes. Documented information is required to be kept (retained) to demonstrate conformity of product and service to requirements and that processes have been carried out as planned. Note that the requirement for processes to accomplish all the operational activities is not repeated in each sub-clause. *A lack of repetitious requirements in each sub-clause does not mean processes and documented information are not required!*

When planning operations, think about the quality objectives (see clause 6.2) and the requirements for the product, which processes and documents need to be established, which specific resources are needed for each product, and which verification, validation, measurement, monitoring, inspection, and test activities are required. The criteria for product acceptance need to be established. The organization also needs to determine which records are required to provide evidence that the realization processes and resulting product meet requirements (see clause 7.5).

Why Do It?

- To achieve customer satisfaction

- To thoroughly understand the processes and activities that will result in products that conform to customer and regulatory/statutory requirements

- To foster continual improvement

- To fully integrate operations for effectiveness and efficiency

Implementation Tips

- Achieve a full understanding of the operational processes and their interactions.

- Map or flowchart the processes necessary to produce conforming products.

- Create quality plans for each product, if appropriate.

- Consider using the product design and development process approach for designing processes.

- Find the vital few key performance indicators for both products and processes; it is better to measure and analyze a few key indicators well than many poorly.

- Align the key process measures with your quality objectives (see clauses 6.2 and 8.2).

- Pay particular attention to determining what documented information you will need as work instructions and evidence of conformance.

Pay particular attention to your intellectual property and consider the need to provide evidence of prudent judgment (that is, records) if liability issues ever arise. Lack of required records or inadequate records could create problems for an organization.

Questions to Ask to Assess Conformity

- Is there evidence of planning of production processes?

- Does the planning extend beyond production processes to encompass all product realization processes?

- Are processes planned, controlled, and operated as outlined in clause 4.4?

- Is the planning consistent with other elements of your QMS?

- Are the operational planning outputs adequate for the organization's needs?

- Is the documented information adequate?

- Do objectives and measures for operational processes align with your quality objectives?

- Are support needs and resources defined during the planning process, and do they appear to be adequate?

- Does the planning define the documented information that must be retained to provide confidence in the conformity of the processes and resulting product?

CLAUSE 8.2 REQUIREMENTS FOR PRODUCTS AND SERVICES

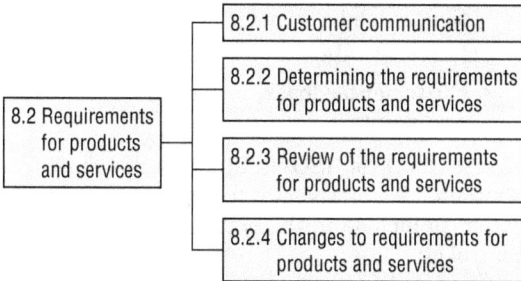

```
                    ┌── 8.2.1 Customer communication
                    │
                    │   8.2.2 Determining the requirements
                    ├──      for products and services
 8.2 Requirements   │
  for products      ┤   8.2.3 Review of the requirements
 and services       ├──      for products and services
                    │
                    │   8.2.4 Changes to requirements for
                    └──      products and services
```

CLAUSE 8.2.1 CUSTOMER COMMUNICATION

What Is the Requirement?

Clause 8.2.1 requires processes to accomplish specific types of information exchange. Similar information exchange requirements are in ISO 9001:2008 clause 7.2.3, which is also titled *Customer communication*. Both versions require that three specific types of communication with customers be included in the organization's processes:

- Product and service information, including customer requirements

- Documented agreements with the customer, such as contracts, orders, changes, and other information that is needed to meet customer requirements

- Customer feedback, including complaints

Clause 8.2.1 of ISO 9001:2015 also has two additional requirements. It requires the organization to include in the organization's processes control of:

- The handling and treatment of customer-owned items, which was covered in great detail in clause 7.5.4 in ISO 9001:2008. (The specific requirements of the 2008 version have been significantly simplified.)

- Any contingency actions that are relevant.

Why Do It?

- To ensure customer satisfaction

- To solve problems quickly and early

- To increase business with customers by "being in touch" on a regular basis

Implementation Tips

- Make the level and form of customer contact consistent with the products you sell and the volume of business with a customer

- Tie the processes needed to meet this requirement to your processes for addressing requirements in clause 9.1.2 relating to monitoring information on customer satisfaction

- Consider processes for contracts or order handling, including changes, and any customer feedback regarding your products or services, including customer complaints

Questions to Ask to Assess Conformity

- Is there evidence of planning of communications with customers?

- Are there processes that cover each type of customer communication that applies in the organization's circumstances?

- Are there effective processes in place to facilitate communication with customers about product information, inquiries, contracts, order handling (including amendments or changes), and customer feedback, including customer complaints?

- Have you linked the processes for communicating with customers with those for monitoring customer satisfaction?

CLAUSE 8.2.2 DETERMINING THE REQUIREMENTS FOR PRODUCTS AND SERVICES

What Is the Requirement?

The organization needs to be sure there is a clear understanding of the requirements specified by the customer (including "soft" requirements for items such as delivery and post-delivery activities), any statutory or regulatory requirements that apply (such as data or records), and any additional requirements included in the QMS that are related to customer orders or quotations.

The organization must also address product requirements that have not been specified by the customer but are necessary for the intended or specified use of the product. This would include "requirements" that are "understood" for a particular product, for example, the careful handling of a package by a package delivery organization.

Why Do It?

- To be sure that the organization can meet customer commitments

- To positively impact customer satisfaction

- To mitigate, minimize, or avoid disputes with customers over what the customer thinks was ordered and what the organization thinks it contracted or committed to provide

- To resolve problems as early as possible

- To minimize the chance of an incomplete quote (for example, missing costs)

Implementation Tips

- Hold face-to-face meetings to resolve issues and anything that is not clear

- Attend pre-bid meetings

- Deploy a process to read quotes and contracts carefully and thoroughly

- Understand "industry standard" and "product standard" unstated expectations

- Keep up to date on regulatory and statutory requirements by subscribing to an update service or by other means

Questions to Ask to Assess Conformity

- Does the organization determine customer requirements?

- Does the process include the determination of requirements needed but not specified?

- Are records available that provide evidence that customer requirements have been determined?

CLAUSE 8.2.3 REVIEW OF THE REQUIREMENTS FOR PRODUCTS AND SERVICES

Clause 8.2.3 contains two untitled sub-clauses, 8.2.3.1 and 8.2.3.2, both of which are addressed below.

What Is the Requirement?

Prior to making a commitment to supply a product to a customer (for example, providing a quote, accepting a contract or order, or accepting a change to an existing contract or order), be sure you understand the customer's requirements. This includes resolving any contract or order requirements that are different from previous agreements or previously issued quotes. Also, be sure that defined requirements (for example, the customer delivery requirement or requirements for post-delivery activities) can be met.

Documented information on the results of any review of customer requirements needs to be maintained.

Where the customer provides no formal statement of requirements, the organization needs to confirm what it intends to supply, via phone, e-mail, or preferably in writing.

When customer requirements change (for example, a change order is issued by a customer), a process is needed to ensure that relevant documents are amended and that personnel involved with an order are made aware of the changed requirements.

Why Do It?

- To ensure that the conditions stated in the order or quote or commitment to a customer can be met

- To provide the products included in the scope of the quotation or tender

- To understand fully and comply with ancillary items such as stated delivery

dates and requirements of applicable
external standards

- To comply with the commercial terms and
  conditions applicable to the order, contract,
  quote, or tender

Implementation Tips

- Balance the risks of noncompliance with
  the effort expended in the review of a
  quotation or a contract.

- Keep in mind that the purpose of the
  review is to add value and not to create a
  bureaucratic review process.

- Have a process for reviewing verbal orders.

- Have a process for off-the-shelf products.

- Have a very simple, brief, and effective
  contract review process for simple
  products.

- Have a more formal process for large,
  complex contracts or quotations,
  which may require the involvement
  of many organizational entities such
  as engineering, manufacturing, legal,
  finance, and quality assurance.

- Consider how to review electronic orders, blanket orders with periodic releases, unsolicited orders, orders through distributors or representatives, faxed orders, Internet orders, and any other type you may receive.

- Consider using quality function deployment (QFD) and developing quality plans where appropriate (see ISO 10005).

If you have unique product requirements that occur only rarely, note in a procedure (that is, documented information) that any such circumstances will be addressed using a specific quality plan to be created when unique occasions arise.

Questions to Ask to Assess Conformity

- Does a process exist that requires the review of identified customer requirements before commitment to supply a product to the customer?

- Does a process exist that requires the review of quotes and orders to ensure that requirements are adequately defined?

- Is there a procedure for handling the review of verbal orders?

- Is there a process documented to handle the resolution of differences between quotations and orders?

- Does a process exist for handling changes to product requirements?

- Is documented information retained of the results of reviews and actions taken?

CLAUSE 8.2.4 CHANGES TO REQUIREMENTS FOR PRODUCTS AND SERVICES

What Is the Requirement?

When changes to product requirements, orders, contracts, or quotations occur, the organization is required to ensure that relevant documented information is amended and communicated, as appropriate, within the organization.

Why Do It?

- To ensure customer satisfaction

- To eliminate provision of nonconforming products

- To solve problems quickly and early

- To increase business with customers by "being in touch" on a regular basis

Implementation Tips

- Consider all avenues for initiation of changes. Changes can come from many different sources. For example, in today's environment floor-level workers often talk directly to factory workers in customers' plants.

- Cell phones are often used to relate the latest changes to schedules and requirements, but the situation can turn into chaos. Control rules are needed so that decisions related to changes are made by the appropriate people with the relevant and up-to-date information.

Often, rapid response is critical for the customer, so design the system in such a way that you can deliver just that!

Questions to Ask to Assess Conformity

- Does a process exist for handling changes to product requirements?

- Are there "hidden" channels for initiating changes that increase your business risk?

4 CLAUSE 8.3 DESIGN AND DEVELOPMENT OF PRODUCTS AND SERVICES

CLAUSE 8.3.1 GENERAL

What Is the Requirement?

This little clause emphasizes that these activities need to be managed as a process.

Why Do It?

- Design and development is often the most important operational activity in the organization.

- Careful management and control of design and development projects can be the key way for the organization to reduce risk.

Implementation Tips

- Manage each design as a separate project.

- Ensure that you have the resources needed—don't try to do too many projects at one time.

- Produce project plans with short intervals between key events to reduce development time and risks.

Questions to Ask to Assess Conformity

- Is there a general design and development process?

- Are individual design and development projects managed separately?

CLAUSE 8.3.2 DESIGN AND DEVELOPMENT PLANNING

What Is the Requirement?

The planning for a design and development project needs to determine design stages considering activities such as verification and validation, control of design interfaces, design review, resources needed for design and development, customer involvement, and the documented information needed to confirm that input requirements are met. Planning is required at the level of detail needed to achieve the design and development objectives—not to generate an excessive amount of paperwork. Stages of the project need to be determined, and responsibilities, authority, and interfaces need to be defined. Requirements need to be established for the incorporation of review, verification, and validation into the design and development project. The organization needs to determine how communications will be structured. In many cases a number of organizations are involved in this process, and the success of the design and development project often rests on proper identification, understanding, and control of design interfaces.

Why Do It?

- To ensure that the product meets customer and regulatory/statutory requirements

- To maximize the probability that projects will be completed on time and within budget

- To manage risk

Implementation Tips

- Generate some form of project flowchart that incorporates the pertinent personnel, timing, and interrelationship information

- Consider approaches such as Gantt charts, PERT (program evaluation and review technique) charts, or CPM (critical path method) charts

- Consider using project management software such as Microsoft Project or Primavera

- Determine the project stages and how the project will proceed from inception to completion

- Plan for the use of failure mode and effects analysis (FMEA) during the design process to prevent problems

Questions to Ask to Assess Conformity

- Are the stages of the design and development project defined? Where?

- Is design review addressed? Where?

- Are verification and validation addressed? Are these activities appropriate?

- Is validation conducted for each product application?

- Is it clear who is responsible for what?

- Is risk, especially liability exposure, considered throughout the design process?

- Are the communication channels defined?
  Is there evidence that communication
  on projects is occurring and that it is
  effective?

CLAUSE 8.3.3 DESIGN AND DEVELOPMENT INPUTS

What Is the Requirement?

This clause requires the organization to determine what input information is pertinent to the product to be designed and developed and to create a requirements specification or an equivalent statement of the general and specific characteristics of the product to be developed, including the suitability of the products and services to meet marketplace and customer needs. Specifically, the organization needs to address:

- Functional and performance requirements, including potential consequences of failures

- Input from customers, when appropriate

- Applicable statutory and regulatory requirements

- Where applicable, information derived from previous similar designs

- Resolution of design conflicts among the inputs

- Potential failure consequences

- Other requirements essential for design and development

- Documented information related to the inputs

Why Do It?

- To maximize the probability that the product will meet defined requirements

- To complete projects on time and within budget

- To reduce the risk of customer dissatisfaction and failure of the products and services

- To minimize the probability of after-delivery detection of service design deficiencies and delivery of nonconforming services

Implementation Tips

Concurrence with the requirements document by all parties is not explicitly required, but it should be considered to avoid misunderstandings during project implementation. It is especially worthwhile to obtain closure, where appropriate, between marketing or sales and those who will be doing the development work.

Items to consider when addressing input information that may be pertinent to the product to be designed and developed include:

- Customer requirements, wants and needs

- Functional requirements

- Performance requirements

- Information from previous similar designs

- Statutory or regulatory requirements

- Environmental considerations such as ISO 14000

- Industry standards

- National and international standards

- Organizational standards

- Safety regulations

- Cost

- Past experiences

- Contract commitments (for designs
  that are related to specific customer
  orders)

- Use of focus groups and QFD

Questions to Ask to Assess Conformity

- Are the requirements for new products
  defined and records maintained?

- Are the requirements complete?

- Are the requirements unambiguous?

- Are the requirements without conflict?
  Is there a process in place to ensure that
  all known conflicts have been resolved?

🔍 CLAUSE 8.3.4 DESIGN AND DEVELOPMENT CONTROLS

What Is the Requirement?

Design and development controls are required.
These include clear delineation of the results
to be achieved, planning and conducting design
and development reviews and verification activities to ensure that design outputs meet input

requirements, and validation to ensure that the products and services meet the requirement for the application intended.

Why Do It?

- To ensure that the products and services will meet requirements and can be produced and delivered to the customer

- To reduce the financial risk of introducing new or updated products and services

Implementation Tips

Achieve a clear understanding of the distinct differences among the three key types of control and execute each appropriately:

- Design and development review is required. This concept applies equally to hardware, processed materials, software, and service projects. In fact, it is a critical element of the software design and development process. When robust design and development reviews are held for software projects, including design and development reviews of software test plans, development cycles are typically reduced and life cycle costs are lower.

Design and development review is intended to address the "abilities" associated with a new product—manufacturability, deliverability, testability, inspectability, shipability, serviceability, repairability, availability, and reliability, as well as issues related to inventory and production planning and the purchase of components and subassemblies. Design and development reviews are intended to identify issues, to discuss possible resolutions, and to determine appropriate follow-up.

- Verification activities are conducted to ensure that the design and development outputs meet the input requirements. Design and development validation is intended to ensure that the design and development output conforms to defined user needs and is capable of meeting the requirements for the specified application or intended use, where known.

- Design and development validation is required for each product and service application. The difference between design and development validation and verification has caused much confusion in the past, especially with new users of the standard. Both verification and validation

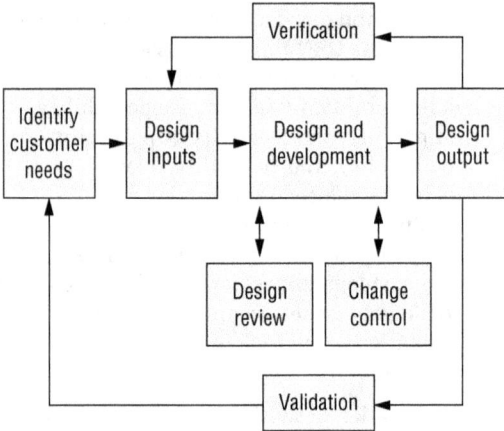

Figure 4.3 Design and development review, verification, and validation.

are explicitly included in ISO 9001:2015 (see Figure 4.3). Definitions of these concepts can be found in ISO 9000: 2015 (see clauses 3.8.12 and 3.8.13).

Questions to Ask to Assess Conformity

- Does the design and development planning include careful consideration of the design and development controls needed—the timing, personnel, equipment, and other resources necessary to carry out the controls?

- Are design and development reviews being performed?

- Are design and development reviews indicated in the project planning documents?

- Who attends design reviews?

- Is the attendance appropriate?

- Are results documented?

- Are follow-up actions taken?

- Are appropriate records maintained?

- Is a verification process in place?

- Is the verification process effectively implemented?

- Is design and development validation performed to confirm that the product is capable of meeting the requirements for intended use?

- Are suitable controls provided in cases where full validation cannot be performed prior to delivery?

- Is documented information (that is, records) of design and development validation maintained?

🔧 CLAUSE 8.3.5 DESIGN AND DEVELOPMENT OUTPUTS

What Is the Requirement?

Design and development outputs are required to meet input requirements, to be adequate for subsequent processes in the provision of the product or service, and to ensure that the products and services are fit for their intended purpose.

This provision of the standard requires that design and development outputs meet defined requirements (as defined in clause 8.3.2), and exist in a form that can be used for subsequent verification. Documented information is required for the output.

Why Do It?

- To provide the information needed to produce the product or deliver the service

- To show that the design and development work has been performed in accordance with requirements

Implementation Tips

- Assign responsibility for documentation of the results of a development project to the person or team performing the work on the project

- Use, as appropriate, mock-ups, models, or other means to communicate the intent of the design and development team

- Use FMEA during the design process to prevent problems

- Maintain development reports or logs that contain data showing that the requirements have been satisfied, where appropriate

- Provide appropriate information to facilitate manufacture of a product to specified requirements

- Use appropriate statistical tools such as design of experiments, hypothesis testing, regression and correlation analysis, simulations, reliability analysis, and statistical tolerancing

- Indicate clear product acceptance criteria in the documentation

- Include any information that relates to producing or using the product safely and properly

- Consider how to demonstrate that the product will not do what it should not do

- Ensure that software or hardware meets requirements and will not interfere with the operation of other software or hardware

- Keep good records (that is, documented information)

- Ensure that the output is approved before a product is released, which is typically achieved by the appropriate management personnel signing off on it

- Consider carefully who should be responsible for deciding on any design releases before completion of design and development review, verification, and validation

Questions to Ask to Assess Conformity

- Is the output of design and development projects in a form suitable for verification against input requirements?

- Does the design and development output satisfy input requirements (for example, as stated in functional requirements specifications)?

- Does output provide, as appropriate, information for production operations?

- Are product acceptance criteria clearly stated?

- Are product safety and product use characteristics identified?

- Is there an approval process for the release of products from the design and development process?

- Are all design and development project changes reviewed to ensure there is no adverse impact on conformity to requirements?

CLAUSE 8.3.6 DESIGN AND DEVELOPMENT CHANGES

What Is the Requirement?

Any changes that occur in the design of products and services, either during the design and development process or after delivery to a customer, need to be identified, and documented information of the changes maintained. Further, changes should be exercised through the design review, verification, and validation processes, and should be approved before implementation.

This clause also now requires evaluation of the effect of changes on constituent parts of the product and on product already delivered.

Why Do It?

- To make certain that changes do not introduce unforeseen adverse effects into the product or other related products or subsystems

- To ensure that design configuration is controlled

- To ensure cost-effective manufacturing and life cycle support, where applicable, after shipment to a customer

Implementation Tips

- Include control of changes that occur during design and development in the document or design control process

- Review changes to ensure that they do not compromise other aspects of the design

- Conduct design and development review, verification, and/or validation in cases where appropriate

- Record and follow up on issues discovered during the review of changes

Questions to Ask to Assess Conformity

- Are design and development changes identified, reviewed, and controlled?

- Are design and development changes design-reviewed, verified, and validated as required?

- Does review of changes include evaluation of their effect on constituent parts and delivered product?

- Is the required documented information on design and development changes maintained?

CLAUSE 8.4 CONTROL OF EXTERNALLY PROVIDED PROCESSES, PRODUCTS AND SERVICES

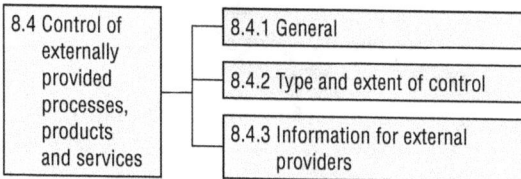

8.4 Control of externally provided processes, products and services	8.4.1 General
	8.4.2 Type and extent of control
	8.4.3 Information for external providers

CLAUSE 8.4.1 GENERAL

What Is the Requirement?

Controls are required to ensure that purchased products and services conform to specified requirements. There is flexibility regarding how this is done. The approach should depend on the effect of the purchased product or service on subsequent product and service processing and delivery operations, on the final product, and on the customers.

Suppliers need to be evaluated and selected based on their ability to supply processes, products, and services in accordance with requirements. Criteria need to be established for selection, evaluation, and reevaluation of external providers. Documented information of the results of these activities and any actions taken as a result of the evaluations must be kept.

Control requirements for externally provided products and services are required to be determined by the organization when:

- The externally provided product or service becomes part of the organization's product or service

- External providers provide product or service directly to the customer on behalf of the organization

- A process or part thereof is outsourced by the organization

Why Do It?

- To ensure that you get what you specify and pay for

- To ensure your own profitability in cases where purchased products and services are a significant component of the cost of goods sold (CGS)

- To ensure that products and services conform to specifications

- To ensure satisfaction of your customers

- To drive robust controls "upstream" as far as possible to ensure the most economical control of quality, minimizing total cost

- To ensure the ongoing ability of your suppliers to continually provide conforming products and services at optimal cost

Implementation Tips

- Understand the risk to your business if purchased products and services do not meet your requirements

- Consider rigorous controls in cases where the potential impact of purchased products and services is great

- Consider reducing the controls in cases where impact of purchased products and services is less

- If a purchased product or service has little impact on the final product, then minimal control is needed

- Think about what makes sense from both a customer and business perspective

- Focus more on obtaining conforming material and not just on maintaining approved-supplier lists

Questions to Ask to Assess Conformity

- Have criteria for the selection and periodic evaluation of suppliers been defined?

- Is there a process for selecting and evaluating suppliers?

- Are the results of evaluations documented and retained as records?

CLAUSE 8.4.2 TYPE AND EXTENT OF CONTROL

What Is the Requirement?

Processes are required to ensure that purchased products and services do not adversely affect the organization's ability to consistently provide conforming products and services. The organization is required to define the controls appropriate to the circumstances. This could be inspection, but inspection is not required. There is flexibility here to choose the most effective approach, taking into consideration the impact of the products or services and the effectiveness of the external provider's controls.

If verification is necessary at a supplier's location, the purchasing information must include this expectation and the requirements for product release.

Why Do It?

- To ensure that material you put into your manufacturing flow meets specified requirements

- To avoid costly rework in subassembly or final assembly or after a product is shipped

Implementation Tips

- Determine controls to be placed on the external provider's processes and on the resulting outputs

- Consider approaches such as:

  - Certifying external providers (based on demonstrated performance, third-party certification, or your own audit)

  - Certifying a specific external provider's product (based on demonstrated process capability)

  - Conventional or skip-lot incoming inspection for attributes or variables using sampling plans

  - One hundred percent inspection (or more)

  - Verification at the supplier's facility

  - Any combination of these or other approaches

- Consider performing verification activities at the supplier's premises if this makes sense

Questions to Ask to Assess Conformity

- Has the organization defined a process for verifying that externally provided products and services conform to defined requirements?

- Is the process effectively implemented?

- Does documented information exist of product acceptance?

CLAUSE 8.4.3 INFORMATION FOR EXTERNAL PROVIDERS

What Is the Requirement?

Applicable requirements shall be provided to external providers for:

- The processes, products, and services to be provided

- Release approval for products and services and any associated process releases

- Any required methods, processes, and equipment

- Competence and qualification of personnel, as appropriate

- Interactions with the organization's quality and management system controls and monitoring to be applied to the external provider

- Intended verification or validation activities at the external provider's location by the organization or its customer

The organization is required to ensure the adequacy of this information before providing it to the outside provider.

Why Do It?

- To maximize the probability that you will receive what you order, on time

- To minimize discussion with suppliers over what you really want

- To forge a partnership mentality with your suppliers

Implementation Tips

- Communicate clearly to suppliers what your organization wants to purchase.

- Communicate clearly to suppliers your criteria for accepting the purchased product.

- Be creative when devising web-based purchasing approaches to ensure adequacy and control of purchase documents.

- Match the process of checking the adequacy of your purchase requirements to the importance of the items. Important items may require that several functions or levels be involved in review and approval (for example, for high-value purchased items); less-important items may require only a single level of review and approval.

Questions to Ask to Assess Conformity

- Is the external provider given adequate information on approval or release of products and services, methods, processes, or equipment, as applicable?

- Is the external provider given adequate information on requirements for personnel competence and qualifications as needed for the application?

- Is the external provider given adequate information for interacting with the organization's QMS and other management systems?

- Is the external provider given adequate information on the control and monitoring to be applied to the external provider's performance?

- Is the external provider given adequate information on verification activities the organization or its customers may perform?

- Do documents include, where applicable, QMS requirements?

CLAUSE 8.5 PRODUCTION AND SERVICE PROVISION

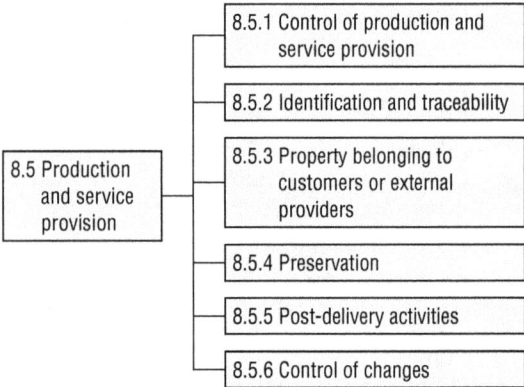

```
                           ┌─────────────────────────────────┐
                        ───│ 8.5.1 Control of production and  │
                           │       service provision          │
                           └─────────────────────────────────┘
                           ┌─────────────────────────────────┐
                        ───│ 8.5.2 Identification and          │
                           │       traceability                │
                           └─────────────────────────────────┘
    ┌──────────────┐       ┌─────────────────────────────────┐
    │ 8.5 Production│      │ 8.5.3 Property belonging to       │
    │ and service   │───  ─│       customers or external       │
    │ provision     │      │       providers                   │
    └──────────────┘       └─────────────────────────────────┘
                           ┌─────────────────────────────────┐
                        ───│ 8.5.4 Preservation                │
                           └─────────────────────────────────┘
                           ┌─────────────────────────────────┐
                        ───│ 8.5.5 Post-delivery activities    │
                           └─────────────────────────────────┘
                           ┌─────────────────────────────────┐
                        ───│ 8.5.6 Control of changes          │
                           └─────────────────────────────────┘
```

These clauses involve actual production of products or delivery of services. They include such activities as making, delivering, and supporting products and services after delivery, traceability, and release of products and services for delivery to the customer.

Clause 8.5 of the new standard is very nearly identical to the requirements of ISO 9001:2008 clause 7.5 of the same title.

🔑 🔢 CLAUSE 8.5.1 CONTROL OF PRODUCTION AND SERVICE PROVISION

What Is the Requirement?

Production and service provision needs to be planned (see clause 8.1) and carried out under controlled conditions. This sentence expresses the essence and intent of ISO 9001:2015. Controlled conditions include, as applicable:

- Having information available that states the characteristics of products or services—such as specifications, drawings, and so on

- Having suitable documented information available where needed—such as procedures, work instructions, specifications, drawings, forms, check sheets, and so on

- Knowing the results to be achieved and criteria of acceptability

- Being able to assess, where possible, that requirements have been fulfilled

- Being able, where appropriate, to take action to address nonconformity

- Knowing where and how to get help to address nonconformity

- Use of appropriate work environment and infrastructure

- Competent and qualified personnel and validated ability to meet requirements

- Taking action to prevent human error

- Knowing what controls are needed and where they are needed, for example, what tests and inspections need to be done and where in the process

- The validation, and periodic revalidation, of the ability to achieve planned results for processes where the output cannot be easily verified (that is, where the product cannot be adequately tested or inspected without making it nonconforming)

- Using suitable equipment to make, monitor, and/or measure the product (or process)

- Knowing and doing all the things that need to be done before product is delivered to a customer or moved to the next operation—for example, reports, sign-offs, and stamps

- Knowing and doing all the things needed to meet both delivery and post-delivery commitments—for example, are there special packaging or shipping instructions, after-shipment start-up service, or ongoing service obligations?

Why Do It?

- To ensure that what you are providing will meet both internal and customer requirements

- To prevent product and process variability that may exceed specification limits or cause high costs, rejects, rework, scrap, and other forms of waste

Implementation Tips

- Understand specifications of products and services

- Identify key processes

- Flowchart or process-map the processes

- Define or clearly identify the outputs of all processes at appropriate stages

- Establish clear criteria of acceptability

- Consider the use of statistical tolerancing

- Identify interrelationships between operations

- Consider preparation of quality plans where appropriate (see ISO 10005)

- Determine what procedures and working instructions are needed for the various processes

- Document processes in a manner suitable to your organization's method of operation

- Make sure process documentation is available when and where needed

- Consider establishing clear criteria for process capability

- Ensure ongoing suitability of equipment by planning maintenance activities

- Plan measuring and monitoring activities in conjunction with planning the controls

- Consider the use of statistical sampling

- Identify devices needed to monitor and measure both the product and its processes

- Make sure that the needed monitoring and measurement equipment is suitable (see clause 7.1.5)

- Consider the use of process capability studies and control charts

- Ensure that planning is compatible with the other processes of the entire interconnected QMS

Questions to Ask to Assess Conformity

- Are specifications available that define quality characteristic requirements of the product or service?

- Has the organization demonstrated the suitability of equipment for production and service operations to meet product or service specifications?

- Has the organization defined all production and service provision activities that require control, including those that need ongoing monitoring, work instructions, or special controls?

- Are work instructions available and adequate to permit control of the appropriate operations so as to ensure conformity of the product or service?

- Have the requirements for the work environment needed to ensure the conformity of the product or service been defined, and are these work environment requirements being met?

- Is suitable monitoring and measurement equipment available when and where necessary to ensure conformity of the product or service?

- Have monitoring and measurement activities been planned, and are they carried out as required?

- For hardware, processed material, and software, have suitable processes been implemented for release of the product and for its delivery to the customer?

- Have suitable release mechanisms been put in place to ensure that products and services conform to requirements?

- Has the organization determined which production or service processes require validation? Have these processes been validated?

- Has the organization defined criteria for the review and approval of production or service processes? Have the reviews and approvals been performed?

- Has the organization determined which personnel need to be qualified, and has it determined the qualification criteria? Have these personnel been qualified?

- Does the organization use defined methodologies and procedures to validate processes?

CLAUSE 8.5.2 IDENTIFICATION AND TRACEABILITY

What Is the Requirement?

The organization needs to identify outputs when necessary to ensure conformity of product and services. Wherever it is appropriate, inspection and test statuses also need to be addressed. When traceability is a requirement (usually because it is included in specifications and/or in a customer order), a method is required to control unique traceability information at all appropriate stages. Documented information shall be retained to enable traceability.

Documented information (that is, records) is required.

Why Do It?

- Identification of product (including material, parts, components, assemblies, and finished goods):

  - Ensures that the status of material will be known at all stages so there is no guesswork about whether a lot, subassembly, or final product has been inspected or tested, or has completed a product realization step

  - Reduces the risk of product failure due to use of the incorrect material, part, or component in production or assembly

- Facilitates easier material control during storage, in-process operations, assembly, and shipment

- Facilitates customers' ability to know that they have received the correct product

- Traceability:

  - Ensures that you can meet any customer requirement to trace a product back to its component elements in case a problem occurs later

  - Is commonly required in industries such as aerospace, automotive, medical devices, oil and gas, and nuclear power for products that are related to health and safety

  - Can be an internal requirement in cases where you wish to reduce the size of potential recalls if problems are found with delivered product

Implementation Tips

- Consider identification by use of traveler cards with lots of materials, bar codes, color codes (especially for machine shop stock), inspection, test, or other stamps, special containers for subassemblies, placement in a specific location, or any other approach that makes sense to the organization.

- Recording of traceability may include heat, lot, batch, or serial numbers.

- Consider using computerized identification systems such as bar coding on product labels or RFID tags that give verification status, item identity, and traceability data.

Questions to Ask to Assess Conformity

- Has the product been identified by suitable means throughout production and service operations?

- Has the status of the product been identified at suitable stages with respect to monitoring and measurement requirements?

- Is traceability a requirement?

- Where traceability is a requirement, is the unique identification of the product recorded and controlled?

CLAUSE 8.5.3 PROPERTY BELONGING TO CUSTOMERS OR EXTERNAL PROVIDERS

What Is the Requirement?

The organization must protect and safeguard property belonging to customers or external providers (for example, a subassembly that will be incorporated into the final product). These controls apply while the property is under control of or being used by the organization. If customer property is lost or damaged, or if it becomes unsuitable for use for any reason, this needs to be documented and reported to the customer.

The standard makes a special point to remind organizations to pay attention to intellectual property of customers to which they have access—software, for example—and to treat such customer property with care.

Why Do It?

- To meet implied obligations to take care of customer property (in addition to all contractual obligations)

- To avoid the time and costs that would be required to resolve any problems with

the customer if you damage, lose, or
misuse customer property

Implementation Tips

- Consider a process to control items such as
  tooling, information, test equipment and
  software, and shipping containers

- Make a brief quality plan specifically
  for the care and handling of customer
  property

Questions to Ask to Assess Conformity

- Has the organization identified, verified,
  protected, and maintained customer
  property that is provided for incorporation
  into the product?

- Does control extend to all customer
  property, including intellectual
  property?

- Does the organization have records that
  indicate when customer property has been
  lost, damaged, or otherwise found to be
  unsuitable?

- Is there evidence that the customer has
  been informed when customer property
  has been lost, damaged, or otherwise

changes to the production and service provision processes?

- Is management review used to share information about upcoming changes?

- Is the control process initiated in a timely manner?

- Is documented information related to changes maintained as required?

CLAUSE 8.6 RELEASE OF PRODUCTS AND SERVICES

What Is the Requirement?

Planned arrangements are required at appropriate stages to verify that products and services conform to requirements. The organization is required to retain documented information of the conformity with acceptance criteria.

Products and services shall not be released to the customer until these planned arrangements have been completed or unless otherwise approved by relevant authority and, if required, by the customer. Documented information is required to be retained as evidence of conformity and traceability to personnel authorizing release.

Why Do It?

- To provide assurance that products continue to meet customer and internal requirements during product production and service delivery

- To guard product and service brand image

- To prevent inadvertent release of nonconforming products and services

- To ensure changes are supported by actions that ensure their success

Implementation Tips

Conformity to requirements needs to be verified, and documented information needs to be retained to substantiate conformity and to indicate who within the organization authorized release of final products and services.

Product release and service delivery require that all specified activities be accomplished unless release is otherwise approved by a relevant authority or by the customer. This usually means that some form of documented information (that is, a record) should be available to document that specified activities have been accomplished.

As appropriate, release methods need to be developed and implemented before providing the product to the customer. Release methods differ in form, timing, and application. For example, airline pilots use preflight checklists to verify that requirements have been met prior to take-off. An automobile repair shop uses both test instruments and a test drive to verify the satisfactory completion of its service before releasing a repaired vehicle to the customer. Also, customer-contact employees can receive immediate feedback by asking customers if the services have been adequately provided.

Questions to Ask to Assess Conformity

- Is there objective evidence that acceptance criteria for product have been met?

- Do records identify the person authorizing release of the product?

- Are all specified activities performed before product release and service delivery?

- If there are instances in which all specified activities have not been performed before product release or service delivery, has a relevant authority (or as appropriate, the customer) been informed and has an authorized individual approved the action?

13   CLAUSE 8.7 CONTROL OF NONCONFORMING OUTPUTS

What Is the Requirement?

Process outputs, products, and services that do not conform to requirements require identification and control. This control is to prevent their unintended use or delivery. The organization is required to deal with the nonconformity by one or more of the following, as applicable:

- Correction of the nonconformity

- Segregation, containment, return, or suspension of provision of products and services

- Informing the customer if and as appropriate

- Obtaining authority to use "as is," under concession

If the nonconformity is subjected to correction, the organization is required to verify final conformity. The organization shall take appropriate corrective action on nonconformities in process outputs, products, and services, including those detected after delivery.

Retention of documented information is required of actions taken and any concessions obtained. This documented information is required to identify the person or authority who decided how to deal with the nonconformity.

Why Do It?

- To prevent unintended use or delivery of product that does not conform to requirements

- To ensure that there is an appropriate review of nonconforming material so that decisions about its disposition are made and documented

Implementation Tips

- Consider how nonconforming product and associated service can be prevented from inadvertent use by identification, segregation, location, or other methods.

- Define who (by job function) has the authority for approval of each type of disposition, for example:

  - A "use as is" disposition might be approved only by the engineering manager (since such a decision is effectively a "change in design" with liability implications).

- – A rework or scrap disposition may
  be approved by manufacturing
  management.

- Keep records of nonconformities, the
  action taken to resolve them, and any
  required internal or external approvals.
  These records may be needed in the future
  (for example, in case of product failure or
  customer dispute, and so on).

Questions to Ask to Assess Conformity

- Is there documented information (that
  is, a procedure) to ensure that product
  and associated service that does not
  conform to requirements is identified
  and controlled to prevent unintended use
  or delivery?

- Is there evidence of appropriate action
  being taken when nonconforming product
  or a nonconforming associated service has
  been detected after delivery or after use
  has started?

- Is it required that any proposed
  rectification of nonconforming product
  be reported for concession to the customer,
  the end user, or a regulatory body?

- Is there objective evidence of appropriate communication with a customer when the organization proposes rectification of nonconforming product?

- Are concessions obtained from customers as appropriate?

CLAUSE 9 PERFORMANCE EVALUATION

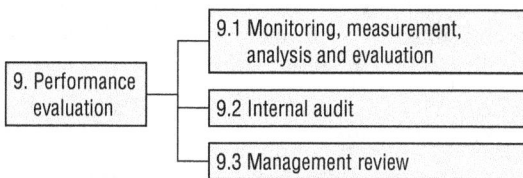

```
9. Performance    ─┬─  9.1 Monitoring, measurement,
evaluation             analysis and evaluation
                   ├─  9.2 Internal audit
                   └─  9.3 Management review
```

CLAUSE 9.1 MONITORING, MEASUREMENT, ANALYSIS AND EVALUATION

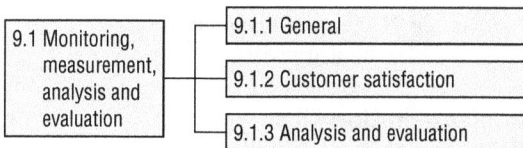

```
9.1 Monitoring,   ─┬─  9.1.1 General
measurement,
analysis and       ├─  9.1.2 Customer satisfaction
evaluation
                   └─  9.1.3 Analysis and evaluation
```

CLAUSE 9.1.1 GENERAL

What Is the Requirement?

The organization must determine what needs to be monitored and measured, how and when this is to be done, and when results are to be analyzed and evaluated.

This includes determination of the methods and techniques to be used to ensure valid results. The organization must also evaluate the performance and effectiveness of the QMS. Appropriate documented information is to be retained as evidence of the results.

Why Do It?

- To ensure that necessary monitoring, measurement, analysis, and improvement activities are planned and implemented

- To ensure the effective implementation of the QMS

- To obtain high return on the investment made in the measurement, analysis, and improvement processes

Implementation Tips

- Think about the processes of the QMS (see clause 4.4)

- Think about the measurement and control needs for operational processes (see clause 8.1)

- Consider monitoring and measurement resources (see clause 7.1.5)

- Use process mapping and flowcharting to integrate measurement, analysis, and improvement activities into operational planning and control

- Decide what you need to know about processes and products to ensure conformity of product and to make certain you meet your quality objectives

- Define the key indicators of performance for the products and processes that will help you understand your progress toward reaching your quality objectives

- Think through all the aspects of measurement, analysis, and improvement activities; keep in mind that you must cover both aspects related to product conformity and aspects related to meeting your quality objectives

- Think about the statistical and other tools that are appropriate for your situation

- Be sure it is clear how you will accomplish the analysis of the data you gather and

Figure 4.4   Monitoring, measurement, analysis, and evaluation requires systems thinking.

how you will use it for improvement of processes

- Consider the interaction of measurement with other processes of the system (see Figure 4.4)

Questions to Ask to Assess Conformity

- Has the organization identified the measurements to be made?

- Is objective evidence available to demonstrate that the organization has defined, planned, and implemented the monitoring and measurement activities needed to ensure conformity and effective QMS implementation and to achieve improvement?

- Is objective evidence available to demonstrate that the organization has determined the need for and use of applicable methodologies, including statistical techniques?

- Does the organization have an ongoing process to determine new measurement needs as product and process changes are developed and implemented?

CLAUSE 9.1.2 CUSTOMER SATISFACTION

What Is the Requirement?

As one of the measurements of the performance of the QMS, it is a requirement to monitor information relating to customer perception as to whether the organization has met customer needs and expectations. There is flexibility for the organization to decide what methods will be used to get this information, but methods must be established.

Why Do It?

- To understand customers' perceptions related to your products so that you can take action to improve that perception

- To understand actions needed to keep current customers

- To understand actions needed to expand your business or increase market share or grow your markets

Implementation Tips

- Decide what information you will monitor, how you will get that information, and how you will use it (see examples of sources of customer satisfaction information in Figures 4.5 and 4.6)

- Consider that to demonstrate conformity in regulated markets, it may be sufficient to monitor customer reports of product deficiencies

- Consider using focus groups and surveys of product users

- If you manufacture products for other companies, consider face-to-face

---

Examples of sources of customer satisfaction information

- Customer complaints
- Returns
- Warranty information
- Customer satisfaction studies
- Results from focus group meetings
- Customer tracking studies
- Questionnaires and surveys
- Reports from consumer organizations
- Direct customer communication
- Benchmarking data
- Industry group information
- Trade association information

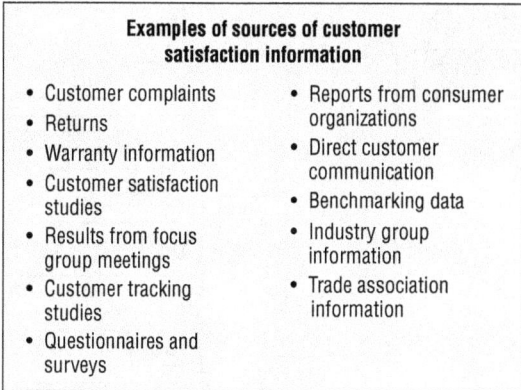

Figure 4.5 Examples of sources of customer satisfaction information.

interviews with the individuals who make key buying decisions in customer organizations.-

- Consider that the products you produce may have associated services that are important to the customers.

- Understand the distinction between monitoring and measurement; monitoring usually provides less information than measuring.

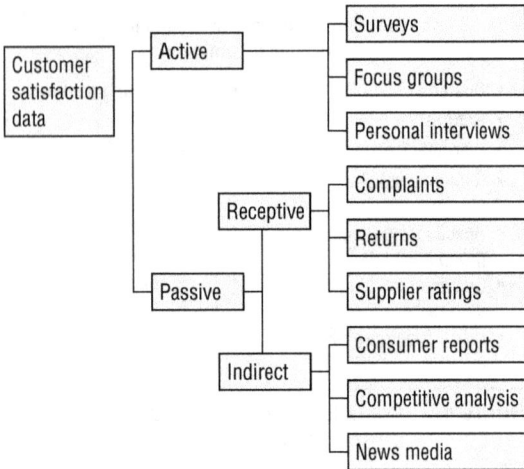

Figure 4.6 Multiple sources of customer data shown in a tree diagram.

- Keep the process dynamic and flexible; the results of monitoring may indicate a need to gather more information through measuring.

- Include methods for understanding what is important to the customers, including price and delivery performance, when deciding what to monitor and measure.

- Consider using QFD and voice of the customer (VOC) techniques.

Questions to Ask to Assess Conformity

- Is customer satisfaction information monitored?

- Are methods for gathering and using customer information determined and deployed throughout the organization?

11 13 CLAUSE 9.1.3 ANALYSIS AND EVALUATION

The organization is required to analyze and evaluate appropriate data and information arising from monitoring and measurement (see Figure 4.7).

The output of the analysis shall be used to evaluate:

- Conformity of products and services to requirements

- Degree of customer satisfaction

- Performance and effectiveness of the QMS

- Whether planning has been effectively implemented

Define how data are used

Figure 4.7 Gathering, analyzing, and using data.

- Effectiveness of actions taken to address risks and opportunities

- The performance of external providers

- The need for improvements to the QMS

Analysis and evaluation of output is also required to be an input to the management review process (see clause 9.3).

Why Do It?

To evaluate:

- The suitability and effectiveness of the QMS

- The degree of customer satisfaction

- Conformity of products and services to requirements

- Performance and effectiveness of the QMS

- The effectiveness of planning

- The effectiveness of actions taken to address risks and opportunities

- The performance of external providers

- The need for improvements to the QMS

Analysis and evaluation of data can also identify improvements that can be made to the QMS to enhance effectiveness and to promote decision making on the basis of fact.

Implementation Tips

- Determine during planning how the data will be used

- Consider the relative importance of the data to be analyzed, and scale the analysis effort to the relative importance of the data

- View the requirements of clauses 5, 6, 7, and 8 as linked in the sense that the organization should function on a closed-loop basis, and data analysis should consider all characteristics of processes

- Use information from the analysis of data as part of the management review process

- Consider the use of "canned" software to alleviate the tedious aspects of data analysis

- Focus data analysis on areas important to achieving the quality objectives; use trend and Pareto charts

- Create an analytical approach to understanding your marketplace and customers

- Understand the information contained in the data; for example:

  - Customers may be delighted even though the product is nonconforming

  - Customers can be highly dissatisfied with product that fully conforms to requirements

  - Identifying the things important to the customer that are causes of dissatisfaction offers an opportunity to change requirements to reflect actual customer needs

  - Understand the effects of variation; consider using histograms and process capability studies

- Consider using SPC, control charts, hypothesis testing, regression analysis, and other statistical tools

One approach that has proven to be effective and efficient for addressing the requirements of clause 9.1 is to first ensure that the key processes required for product realization, support, and improvement are identified. We recommend careful attention to clauses 8.1 and 8.5, which underscore the requirements for planning and controlling processes. For each key process, the inputs to and outputs from the process can be defined. Once the outputs of key processes are defined and understood, it is possible to determine how to measure or monitor the outputs to ensure that they meet requirements. While considering outputs and how to measure or monitor them, it is advisable to also consider and provide evidence of any documented information (that is, records) that may be required to ensure conformity to requirements.

Such an approach can be implemented using a template consisting of a 3 × 2 matrix of blocks. The upper-left block can be used for listing the process inputs, the top middle block for listing the activity characteristics, and the upper-right block for listing the process outputs. The lower-left block can contain the interfaces or interactions with the activity being analyzed, the lower middle block a list of the records required, and the lower-right block a list of the metrics to be used to ensure that the output meets requirements.

A simplified example of such a template completed for the corrective action (CA) process (see clause 10.2) could look like Figure 4.8.

Questions to Ask to Assess Conformity

- Has the organization determined the appropriate data to be collected?

- Does the organization analyze the appropriate data to determine the suitability and effectiveness of the QMS?

- Does the organization analyze the appropriate data to identify improvements that can be made?

- Does the organization analyze the appropriate data to provide information on customer satisfaction?

- Does the organization analyze the appropriate data to provide information on conformance to product requirements?

- Does the organization analyze the appropriate data to provide information on characteristics of processes, products, and their trends?

Inputs
- Undesirable process or condition
- Customer complaint

Process
- Review request
- Proceed with corrective action
- If yes, develop CA plan
- Implement plan
- Check effectiveness

Outputs
- Implemented CA
- Documentation to management review file

Interactions
- Depends on CA; could be purchasing department, design engineering, training department, and so on

Records
- CA plan
- CA effectiveness

Metrics
- CA log
- CA closeout
- Time to closeout

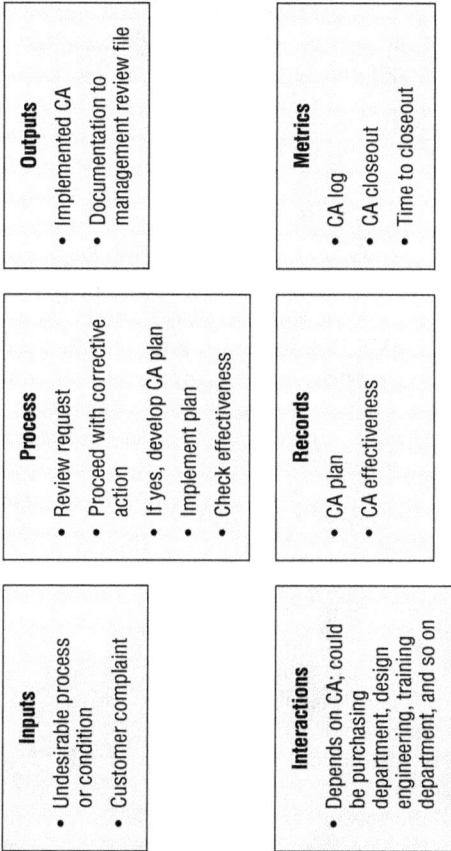

Figure 4.8 Example of a template for the corrective action process.

- Does the organization analyze the appropriate data to provide information on suppliers?

CLAUSE 9.2 INTERNAL AUDIT

What Is the Requirement?

Internal audits are required at planned intervals to determine whether the QMS conforms to the requirements of the ISO 9001:2015 standard and to its own QMS requirements, and to determine if it is effectively implemented and maintained.

Internal audit planning needs to consider the status and importance of the processes and areas to be audited as well as the results of previous audits. The audit criteria, scope, frequency, and methods all need to be defined.

The selection of auditors and conduct of audits needs to be done on a basis that ensures objectivity and impartiality of the audit process. Auditors cannot audit their own work.

A documented procedure is required to describe the responsibilities and requirements for planning and conducting audits, and for reporting results and maintaining records. The organization is also required to retain documented information as evidence of the implementation of the audit program and the audit results.

The organization is responsible for ensuring that actions are taken without undue delay to eliminate detected nonconformities and their causes. Follow-up activities need to include the verification of the actions taken and the reporting of verification results.

Why Do It?

- To provide confidence in the effective implementation of the QMS

- To identify opportunities for improvement in addition to assessing compliance

- To keep everyone sharply focused on adhering to processes, minimizing the natural tendency of "conformity drift"

Implementation Tips

- Use ISO 19011 as a guide to set up your audit processes; choose from its many guidance items

- Consider assigning organization, basic planning, and documentation integrity for the internal audit process to one area, such as the internal audit department or the quality assurance department

- Consider combining and integrating common aspects of management system audits for quality, health and safety, and the environment

- When evaluating the QMS, the following questions are typical of what should be considered for every process being evaluated during internal audits:

  - Is the process identified and appropriately described?

  - Are responsibilities assigned?

  - Are required processes implemented, controlled, maintained, and improved?

  - Are processes monitored and measured as appropriate?

  - Are processes operated under controlled conditions?

  - Are processes effective in achieving the required results?

- Ensure that internal audits consider the processes of the QMS, not just elements or activities

- Select internal auditors carefully (individuals from areas such as finance, engineering, or top management often make excellent quality auditors) and provide good training for them

- Consider both the status and importance of areas and the results of prior audits when determining audit frequency; audit the most critical areas most often

- Consider using the audit process to find opportunities for improvement (see Arter, Cianfrani, and West, *How to Audit the Process-Based QMS* [2012] for more details on auditing processes)

Questions to Ask to Assess Conformity

- Does the organization conduct periodic audits of the QMS?

- Do the periodic audits evaluate the conformity of the QMS to the requirements of ISO 9001:2015?

- Do the periodic audits evaluate the degree to which the QMS has been effectively implemented and maintained?

- Does the organization plan the audit program with regard to the status and importance of areas to be audited?

- Does the organization plan the audit program with regard to the results of previous audits?

- Are the audit scope, frequency, and methodology defined?

- Do the audit process and auditor assignment ensure objectivity and impartiality?

- Is there a documented procedure that includes the responsibilities and requirements for planning and conducting audits?

- Is there a process for ensuring the objectivity and impartiality of audits and auditors?

- Is timely corrective action taken on deficiencies found during the audit?

- Do follow-up actions include the verification of the implementation of corrective action?

- Do follow-up actions include the reporting of verification results?

2 CLAUSE 9.3 MANAGEMENT REVIEW

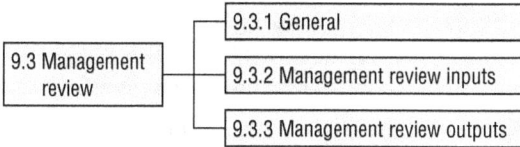

```
                        ┌─ 9.3.1 General
 9.3 Management ────────┼─ 9.3.2 Management review inputs
    review              │
                        └─ 9.3.3 Management review outputs
```

CLAUSE 9.3.1 GENERAL

What Is the Requirement?

Top management is required to review the QMS at planned intervals. This review is to determine the suitability, adequacy, and effectiveness of the QMS. The review needs to be planned to include assessment of the need for changes to the quality policy, the quality objectives, and the processes of the QMS. The review process also needs to include verification that the QMS is aligned with the strategic direction of the organization.

Why Do It?

- To ensure the continuing suitability of the QMS to meet its current purpose and the quality policy

- To ensure the continuing adequacy of the QMS in terms of its breadth and depth of coverage

- To ensure the continuing effectiveness of the QMS in meeting the quality objectives and carrying out planned activities

- To discover issues that require changes to the system to improve its effectiveness

- To determine opportunities for improvement of the QMS and its processes

- To provide necessary resources

- To reallocate resources as changes to the QMS occur

- To remove roadblocks to improvement

- To make prioritization decisions

Implementation Tips

- Be sure management review is a process, not just a meeting.

- Limit meetings to X hours ($X < 2$ is recommended).

- Be sure your top managers do the review personally; it is not a responsibility that can be delegated.

- Use an annual planning session to review the past year, set next year's objectives, and make resource allocations.

- Hold management review meetings monthly or quarterly to review progress, decide on actions, and reallocate resources.

- Be sure preparation by your staff is done competently.

- Make the review meetings useful to top managers by focusing on achieving results and characterizing issues in management language (for example, $$$).

- A few characteristics of successful
  management reviews are things like:

  - A top management *attitude* that the
    system can and should be improved.
    The objective of management review is
    to determine how. Management reviews
    are not celebrations of what went right;
    they are reviews to determine how to
    make more things go great!

  - *Preparation!* It is essential. Someone
    needs to get all the inputs together and
    make sense of them before the review
    starts. This is so top managers can see
    issues from all relevant perspectives.
    For example, aggregation of internal
    process data with customer complaint
    information can paint a clear picture
    of an opportunity to improve customer
    satisfaction while the improvement
    process also reduces costs.

  - *Follow-up* is necessary to make certain
    that decisions are clear and action is
    taken. Frequent and formal progress
    reviews are needed. Robust, effective
    management reviews can be developed
    by starting them at the beginning of
    QMS implementation. Starting the
    process early can facilitate systematic
    improvement of the review process over

the implementation phase so that it becomes a key element of organizational success.

– *Having and following a crisp agenda.* The reviews are not problem-solving meetings but rather an opportunity to assess status and make decisions regarding who is going to do what and by when.

Questions to Ask to Assess Conformity

- Does top management review the QMS at planned intervals to ensure its continuing suitability, adequacy, and effectiveness?

- Do the management reviews include evaluation of the need for changes to the organization's QMS, including the quality policy and quality objectives?

5 6 7 11 CLAUSE 9.3.2 MANAGEMENT REVIEW INPUTS

What Is the Requirement?

Inputs for the management reviews need to include customer feedback and feedback from relevant interested parties. Consideration is to be given to effectiveness of actions on risks and opportunities. The adequacy and allocation of resources needs consideration. The reviews need to include information and trends related to:

- Customer satisfaction
- Process and product performance
- Nonconformities and corrective actions
- Performance of external providers
- Changes that could affect the QMS
- Results of audits
- The extent to which quality objectives have been met
- Opportunities for improvement

In addition, follow-up actions from earlier management reviews need to be included.

Why Do It?

- To make the management review efficient and effective
- To focus the review on important issues
- To provide data with which top managers can make objective decisions and set priorities

Implementation Tips

- Consider including additional inputs beyond the minimum requirements

- Have staff members provide the inputs; in small organizations, data may be collected and provided by top managers themselves

- Use appropriate, simple charts to present trends in quantitative data

- Provide the inputs to attendees prior to the review meetings where feasible

- Use a tool such as a Pareto chart to focus the review on those inputs that require top managers to make decisions, implement change, or provide resources

- When there is a need to focus on variation, display data using histograms or other appropriate tools

Questions to Ask to Assess Conformity

- Does management review input include results of audits, customer feedback, process performance, product conformity, status of prior actions, follow-up actions from earlier management reviews, and changes that could affect the QMS?

- Does management review input include analysis to focus the discussion on areas needing improvement?

- Are recommendations for improvement, actions needed, and resources included in the reviews?

CLAUSE 9.3.3 MANAGEMENT REVIEW OUTPUTS

What Is the Requirement?

Management review outputs include actions to be taken and decisions made for improving the QMS and its processes. Top managers need to allocate or reallocate resources to make the improvements happen.

Why Do It?

- To make the management review efficient and effective

- To focus the review on important issues

- To provide data with which top managers can make objective decisions

- To focus on action and getting results

Implementation Tips

- Consider actions that focus on process improvements, including actions to eliminate waste, to simplify or foolproof processes, to develop improved methods, to improve documented information, and so on

- Include actions to address valid customer complaints, field failures, and other failures to meet customer expectations

- Since customer needs and expectations may change often, organizations may want to consider anticipating new customer requirements and implementing actions to address them

- Ensure that resources are provided as needed for continual operation and improvement of the QMS

Questions to Ask to Assess Conformity

- Do the outputs of management reviews include actions related to the improvement of the QMS and its processes?

- Do the outputs of management reviews include actions related to the improvement of product that does not meet customer requirements?

- Do the outputs of management reviews address resource needs?

- Is documented information retained as evidence of the results of management review inputs?

🕚   CLAUSE 10 IMPROVEMENT

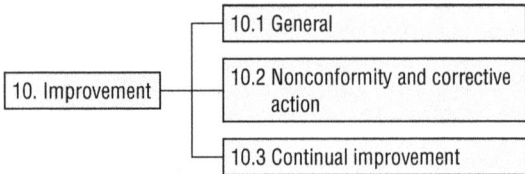

10. Improvement	10.1 General
	10.2 Nonconformity and corrective action
	10.3 Continual improvement

CLAUSE 10.1 GENERAL

What Is the Requirement?

Clause 10.1 requires organizations to determine and select opportunities for improvement and implement actions necessary to meet customer requirements and enhance customer satisfaction.

The opportunities for improvement stated for inclusion, as appropriate, are:

- Correcting, preventing, or reducing undesired effects

- Improving products and services to meet requirements as well as to address future needs and expectations

- Improving QMS performance and effectiveness

A Note to this clause indicates that improvement initiatives can be reactive (for example, corrective action), incremental (for example, continual improvement), by step change (for example, breakthrough), creative (for example, innovation), or by reorganization (for example, transformation). The Note also includes the notion of "correction" as an example of improvement.

Why Do It?

- To become more competitive

- To survive

- To operate more effectively

- To improve the ability to meet customer requirements

Implementation Tips

- During development and deployment of processes of the QMS, keep the notion of improvement at the forefront of your work.

- Address the improvement requirements during planning of the QMS.

- Consider competitive benchmarking.

- Work on improving system and process effectiveness and efficiency at the same time.

- Consider aggressive improvements in *every* element of the QMS.

- Any aspect, process, or element of the QMS that is not important to the organization's success should be identified and targeted for elimination or combination with other elements.

- When implementing improvement projects, care must be taken to consider process interactions.

- Strive to achieve an integrated improvement mentality throughout the organization.

- Prioritize improving what is important for the organization's long-term success.

- Educate the organization on the difference between correction and corrective action.

3 5
6 7
8 12
13

CLAUSE 10.2 NONCONFORMITY AND CORRECTIVE ACTION

CLAUSE 10.2.1 (UNTITLED CLAUSE)

What Is the Requirement?

The organization is required to react to nonconformities, including those arising from complaints, and, as applicable, take action to control and correct them and to deal with the consequences.

The need for action to eliminate the cause(s) of the nonconformity must be evaluated to ensure that it does not recur or occur elsewhere. This means reviewing the nonconformity, determining its causes, and determining whether similar nonconformities exist, or if conditions exist under which similar nonconformities may occur.

The organization is also required to review the effectiveness of any corrective action taken, and to make changes to the QMS as appropriate. Although the standard does not specifically state what actions shall be taken, organizations are encouraged to not only aggressively pursue root cause analysis and corrective action but also address any related training and documentation that are appropriate, as well as schedule future internal audits to ensure that effective corrective action implementation is sustained. Corrective actions, it is noted, shall be appropriate to the effects of the nonconformities that are encountered.

Why Do It?

- To become more competitive

- To survive

- To prevent recurrence of nonconformities and other problems

- To improve the ability to meet customer requirements

Implementation Tips

- Focus on the most important nonconformities first.

- When repetitive nonconformities are incurred, focus on finding process changes that can reduce their frequency of occurrence.

- Addressing and resolving nonconformity is best achieved by having documented information (that is, a procedure) that details how the organization shall react when nonconformity occurs. The process needs to address several activities, such as:

  - How to deal with the consequences of the nonconformity

  - How to initiate action to control the nonconformity

  - How to perform root cause analysis

  - Determining the causes of the nonconformity

  - How to decide on a course of action

  - Determining whether similar nonconformity can exist elsewhere

  - Initiating any actions needed

  - Reviewing the effectiveness of the actions taken

  - Making any changes to the QMS

   – Institutionalizing corrective action
     (for example, training, documented
     information improvements, auditing)

Deploying such a process will ensure consistent
handling of nonconformity. It is also important to
ensure that everyone in the organization under-
stands that there is a distinct difference between
correction and corrective action.

CLAUSE 10.2.2 (UNTITLED CLAUSE)

What Is the Requirement?

The organization shall retain documented information as evidence of the nature of the nonconformities and any subsequent actions taken and
the results of any corrective actions.

Why Do It?

   • To provide input for further analysis if
     nonconformities do recur

   • To ensure that action is taken, even if
     the problem is thorny and the cause
     proves elusive

   • To survive

Implementation Tips

- Define and document a process (that is, a procedure) for addressing nonconformity, correction of nonconformity, and corrective action. It is unfortunate that in many organizations the difference between correction and corrective action is not understood or is ignored.

- Develop a process that requires true corrective action (that is, identification and either elimination or at least minimization of root causes of nonconformity).

- Drive into the workforce an understanding of the correct meaning of corrective action.

- Remember that, in some cases, action may be neither required nor appropriate.

CLAUSE 10.3 CONTINUAL IMPROVEMENT

What Is the Requirement?

The organization shall continually improve the suitability, adequacy, and effectiveness of the QMS.

The organization shall consider the outputs of analysis and evaluation, and the outputs from management review, to decide whether areas of underperformance or opportunities shall be addressed as part of continual improvement.

Why Do It?

- To become more competitive

- To survive

- To operate more effectively

- To improve the ability to meet customer requirements

Implementation Tips

- Set quality objectives with improvement in mind

- Ensure that top management uses management reviews to identify opportunities to improve the QMS and its processes

- Incorporate an improvement mentality and risk-based thinking into the internal audit process

- Use corrective action and associated problem-solving processes as tools for improvement of processes

- Some specific items to consider including in an improvement process are as follows:

  - Performing a failure mode, effects, and criticality analysis (FMECA) in the design stage

  - Performing reliability analyses to make reliability trade-offs (for example, mean time between failures, Weibull modeling)

  - Fault tree analysis

  - Data analysis of other areas of the organization to identify where similar issues may exist

  - Corrective actions to determine whether there is potential to expand the scope to embrace improvement and risk aversion

  - Consideration of analytics employed in Six Sigma, lean, and other similar improvement methodologies

- Train individuals on the use of contemporary tools

Questions to Ask to Assess Conformity

- How does the organization determine and select opportunities for the continual improvement of the quality management system?

- How does the organization implement necessary actions to meet customer requirements and enhance customer satisfaction?

- Does the organization consider selection of opportunities to improve processes to prevent nonconformities?

- Does the organization consider selection of opportunities to improve products and services to meet known and predicted requirements?

- Does the organization consider selection of opportunities to improve quality management system results?

- Does the organization take corrective action to control and correct nonconformities?

- How does the organization address the consequences of nonconformity?

- Does the organization evaluate the need for action to eliminate the cause(s) of the nonconformity to ensure that it does not recur or occur elsewhere?

- Does it review the nonconformity, determine the causes of the nonconformity, and determine whether similar nonconformities exist, or could potentially occur?

- How does the organization implement any action needed?

- How does the organization review the effectiveness of any corrective action taken?

- How does the organization make changes to the quality management system, if necessary?

- Does the organization retain documented information as evidence of the nature of the nonconformities and any subsequent actions taken and the results of any corrective action?

- Is the corrective action taken appropriate to the impact of the problems encountered?

- How is the organization continually improving the suitability, adequacy, and effectiveness of the quality management system?

- How is the organization considering the outputs of analysis and evaluation, and the outputs from management review, to confirm whether there are areas of underperformance or opportunities that shall be addressed as part of continual improvement?

- How is the organization selecting and utilizing applicable tools and methodologies for investigation of the causes of underperformance and for supporting continual improvement?

ANNEX A: CLARIFICATION OF NEW STRUCTURE, TERMINOLOGY AND CONCEPTS

This annex is included in the standard to provide clarification and explanation related to some of the content of the standard. Inclusion of this annex should provide the astute user of the standard with the insight that the standard requires careful reading to ensure implementation is addressed in a way that both meets requirements and provides value to the organization.

A.1 STRUCTURE AND TERMINOLOGY

To improve alignment with other management systems standards (MSSs) the titles and numbering of the clauses and some of the terms in ISO 9001:2015 have been changed from the previous edition (ISO 9001:2008). This change is intended to enhance alignment with other management systems standards.

It is important for organizations to note that *Annex A of ISO 9001:2015 makes a specific point that there is no requirement for an organization to apply the structure and terminology of the standard to the quality management system of an organization.* Rather than being a model for documenting an organization's policies, objectives, and processes, the structure of clauses is intended to provide what the drafters of the standard believed to be a coherent presentation of structure and requirements within the restraints imposed by conforming to Annex SL.

It is also important to note that *Annex A states that there is no requirement for the terms used by an organization to be replaced by the terms used in this standard to specify requirements.* Organizations can choose to use terms that suit their operations (for example, using "records," "procedures," "documentation," or "protocols" rather than "documented information," or "supplier," "partner," or "vendor" rather than "external provider"). Table 4.3 shows the major differences in terminology between ISO 9001:2015 and ISO 9001:2008.

Table 4.3 Major differences in terminology between ISO 9001:2008 and ISO 9001:2015.

ISO 9001:2008	ISO 9001:2015
Products	Products and services
Exclusions	Not used (See clause A.5 for clarification of applicability)
Management representative	Not used (Similar responsibilities and authorities are assigned but no requirement for a single management representative)
Documentation, quality manual, documented procedures, records	Documented information
Work environment	Environment for the operation of processes
Monitoring and measuring equipment	Monitoring and measuring resources
Purchased product	Externally provided products and services
Supplier	External provider

A.2 PRODUCTS AND SERVICES

After 29 years of adhering to a formal definition of *product* to include service (that is, product is hardware, software, processed materials, or services, or any combination of these) ISO dictated that product be redefined as ". . . output of an organization that can be produced without any transaction taking place between the organization and the customer . . ." along with three Notes that can either clarify or obfuscate the meaning of the word. Various rationales have been offered for this change, none of which are credible in the opinion of the authors.

This edition of the standard uses "products and services" throughout, which we believe is intended to include all output categories (hardware, services, software, and processed materials).

A.3 UNDERSTANDING THE NEEDS AND EXPECTATIONS OF INTERESTED PARTIES

The requirement in clause 4.2 is for the organization to determine the interested parties that are relevant to the quality management system. This requirement can be confusing. The scope clause states that ". . . the standard is applicable where an organization needs to demonstrate its ability to consistently provide products and services that meet customer and applicable statutory and regulatory requirements . . ." The scope says nothing about interested parties.

We believe an organization should give this requirement careful consideration since "interested parties" beyond the customer (society, suppliers, workers, regulators, and so on) can impact sustainability.

There is certainly no requirement for the organization to consider interested parties where it has decided that those parties are not relevant to its objectives and to its QMS. But we do suggest that the organization give this requirement some thought, perhaps even develop and implement a process to decide if there are interested parties relevant to its QMS, and if so, what their requirements are.

A.4 RISK-BASED THINKING

The concept of risk-based thinking has been implicit in ISO 9001 through requirements for planning, internal audit, design review, management review, and continual improvement. The latest edition specifies requirements for the organization to understand its context (in clause 4.1) and determine risks as a basis for planning (in clause 6.1). This represents the application of risk-based thinking to planning and implementing QMS processes (see clause 4.4), and will assist in determining the extent of documented information that the organization may require to prevent risks and nonconformities.

Organizations should recall when planning their QMS that there is no requirement for formal methods for risk management or a documented risk management process. Organizations can decide what risk management methodology is appropriate. What is required under clause 6.1 is the application of risk-based thinking and the actions it takes to address risk, including whether or not to retain documented information as evidence of its determination of risk.

A.5 APPLICABILITY

In the past there were provisions for excluding the applicability of requirements from the QMS. Now the standard does not refer to "exclusions."

Now clause 4.3 defines conditions under which an organization can decide that a requirement cannot be applied to any of the processes within the scope of its QMS. The organization can only decide that a requirement is not applicable if its decision will have no impact on its ability to achieve conformity of products and services to requirements.

A.6 DOCUMENTED INFORMATION

Annex SL dictated the adoption of a common clause on "documented information" without significant change or addition to the SL text (see clause 7.5). Where appropriate, text elsewhere in the standard has been aligned with this requirement and the "documented information" terminology is used for all document requirements.

Where ISO 9001:2008 used terminology such as "document" or "documented procedures," "quality manual" or "quality plan," this edition of the standard defines requirements to "maintain documented information."

Where ISO 9001:2008 used the term "records" to denote documents needed to provide evidence of conformity with requirements, this is now expressed as a requirement to "retain documented information." The organization is responsible for determining what documented information needs to be retained, the period of time for which it is to be retained, and the media to be used for its retention.

A requirement to "maintain" documented information does not exclude the possibility that the organization might also need to "retain" that same documented information for a particular purpose, for example, previous versions of it for later reference.

Where the ISO 9001:2015 standard refers to "information" rather than "documented information" (for example, in clause 4.1, "The organization shall monitor and review the information about these external and internal issues"), there is no requirement that this information is to be documented. In such situations, the organization can decide whether or not it is necessary or appropriate to maintain documented information.

A.7 ORGANIZATIONAL KNOWLEDGE

Clause 7.1.6 addresses a requirement to determine and manage the knowledge maintained by the organization to ensure that it can achieve conformity of products and services.

Requirements regarding organizational knowledge were introduced to ensure safeguarding of knowledge and preventing its loss through staff turnover or retirement, to encourage anticipation of future knowledge needs that may have a long time constant to acquire, and to retain tacit knowledge that may not be documented.

A.8 CONTROL OF EXTERNALLY PROVIDED PROCESSES, PRODUCTS AND SERVICES

All forms of externally provided products and services are addressed in clause 8.4, including products and services that are purchased, acquired from an affiliated organization (for example, another division of the organization), or outsourced to an external provider.

The controls required for an external provider of products and services can vary widely. Risk-based thinking can be utilized to determine the type and extent of the controls that are appropriate to be applied to externally provided products and services.

ANNEX B (INFORMATIVE) OTHER INTERNATIONAL STANDARDS ON QUALITY MANAGEMENT AND QUALITY MANAGEMENT SYSTEMS DEVELOPED BY ISO/TC 176

ISO/TC 176 has developed and published many standards to provide supporting information and to provide guidance for organizations that choose to both implement QMSs as well as progress beyond meeting the minimum QMS requirements contained in ISO 9001. Requirements contained in the documents listed in this annex do not add to or modify the requirements of ISO 9001.

Annex B of ISO 9001:2015 contains information on the following standards that can be of assistance to organizations when they are establishing or seeking to improve their quality management systems, their processes, or their activities. Further information on the following standards can be obtained from ISO or ASQ:

- ISO 9000 *Quality management systems— Fundamentals and vocabulary*

- ISO 9004 *Managing for the sustained success of an organization—A quality management approach*

- ISO 10001 *Quality management— Customer satisfaction—Guidelines for codes of conduct for organizations*

- ISO 10002 *Quality management— Customer satisfaction—Guidelines for complaints handling in organizations*

- ISO 10003 *Quality management— Customer satisfaction—Guidelines for dispute resolution external to organizations*

- ISO 10004 *Quality management—Customer satisfaction—Guidelines for monitoring and measuring*

- ISO 10005 *Quality management systems—Guidelines for quality plans*

- ISO 10006 *Quality management systems—Guidelines for quality management in projects*

- ISO 10007 *Quality management systems—Guidelines for configuration management*

- ISO 10008 *Quality management—Customer satisfaction—Guidelines for business-to-consumer electronic commerce transactions*

- ISO 10012 *Measurement management systems—Requirements for measurement processes and measuring equipment*

- ISO/TR 10013 *Guidelines for quality management system documentation*

- ISO 10014 *Quality management—Guidelines for realizing financial and economic benefits*

- ISO 10015 *Quality management—Guidelines for training*

- ISO/TR 10017 *Guidance on statistical techniques for ISO 9001:2000*

- ISO 10018 *Quality management—Guidelines on people involvement and competence*

- ISO 10019 *Guidelines for the selection of quality management system consultants and use of their services*

- ISO 19011 *Guidelines for auditing management systems*

Chapter 5

Tools

TOOLS DISCUSSED IN THIS CHAPTER

This chapter outlines a limited number of tools commonly used in the application of ISO 9001:2015 requirements to manufacturing processes. The amount of detail included in the description of each tool is intentionally limited. For most of the tools included, entire books are available that describe the tools and their application in great detail. Our intent is to provide an overview of:

- What is it?

- Where is it used?

- How is it done?

- Cautions to be considered when using the tool.

Where appropriate, examples of the use of the tool are provided. The user should consult appropriate texts for more-detailed information to ensure correct application in specific situations.

The following tools are included in this section:

Tool number	Tool	Used with ISO 9001:2015 clauses
1	Flowchart	4.1, 4.4, 6.2, 6.3, 8.1, 8.3.2, 8.4, 8.5.1, 9.1.1
2	Process mapping	4.1, 4.4, 6, 7.1, 7.5, 8.5, 9.1, 9.3
3	Brainstorming	4.1, 4.4, 6.2, 10.2
4	Gantt charts	5.3, 7.1.1, 8.3
5	Run or trend charts	8.5, 9.3.2, 10.2
6	Histograms	9.3.2, 10.2
7	Pareto charts	9.3.2, 10.2
8	Failure mode and effects analysis	6.1, 8.3.2, 8.3.5, 10.2
9	Reliability analysis	8.3.2, 8.3.4

Continued

Tool number	Tool	Used with ISO 9001:2015 clauses
10	Identifying external and internal issues	4.1, 4.2, 4.3, 4.4, 6.1
11	Process capability studies	8.5.1, 8.5.6, 9.1.3, 9.3.2, 10
12	Cause-and-effect diagrams	10.2
13	Problem solving	8.5, 8.7, 9.1.3, 10.2
14	How to conduct an improvement project	All clauses
ALL	All tools	

TOOL 1: FLOWCHART

What Is It?

A *flowchart* is a picture of the actual flow or sequence of events that occurs in a process. Flowcharts can be at a "high level," showing only the major elements of a process or system. They can also be very detailed, for example, showing the specific steps in manufacturing a subassembly.

An accurate flowchart can also be used to identify opportunities for improvements in processes.

Flowcharts typically use the following symbols:

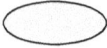

Oval—begins or ends the process

Rectangle—activity or a step in the process

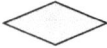

Diamond—decision point

Wait or delay symbol—hold or wait point in the process

Flow line or arrow—shows the direction of flow

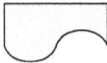

Document—shows the need to create a document or record

Where Is It Used?

Anywhere that processes exist, flowcharts can be considered as a tool to understand the elements of the processes. Flowcharts can be used to:

- Identify the interactions of the overall processes of the QMS

- Represent the steps in the operational processes of an organization

- Aid in developing plans for monitoring and measuring.

How Is It Done?

- Decide on the process to be flowcharted
- Identify the steps of the process (for example, use brainstorming)
- Place the steps in order
- Draw the flowchart using the appropriate symbols (as given above)
- Consider using the tools in Microsoft Word or PowerPoint to create flowcharts
- Connect the steps with arrows
- Verify the accuracy of the flowchart
- Validate the flowchart with individuals familiar with the process

See example below of a flowchart for a corrective action process.

Cautions

Flowcharts can be made so complex (or so simple) that they are useless. Care must be exercised when selecting the scope of the process to be flowcharted—not too simple, not too complex, not too much detail, not too little. Also, it is beneficial to involve individuals with intimate knowledge of the process being flowcharted to ensure that an accurate picture is created.

System-level corrective action

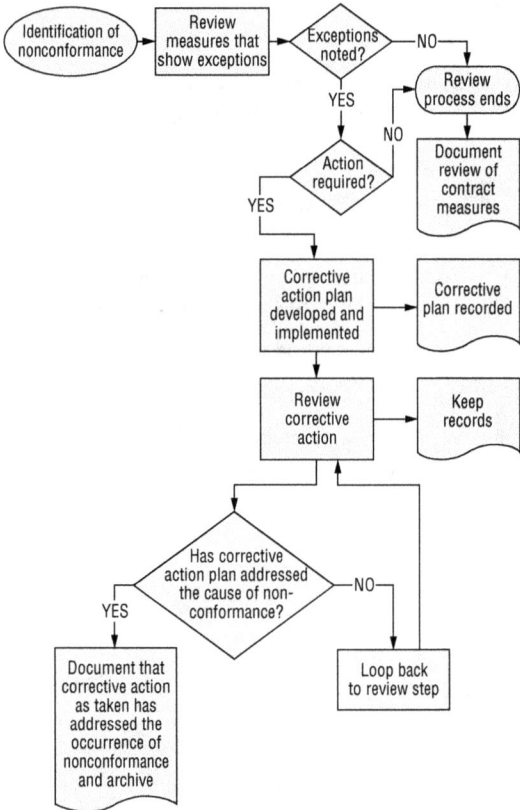

TOOL 2: PROCESS MAPPING

What Is It?

A *process map* describes a process in detail, considering the outputs from the process and the inputs to the process. It also visually displays the value-adding steps in the process that convert the inputs into the desired outputs. Process mapping is the flowcharting of a work process in detail, including key measurements.

Process mapping
Flowchart + Measures

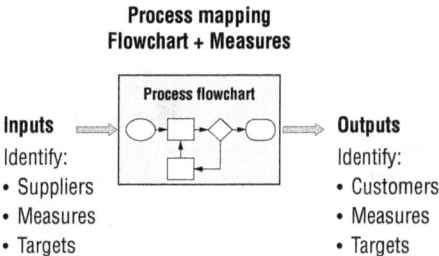

Inputs
Identify:
• Suppliers
• Measures
• Targets

Process flowchart

Outputs
Identify:
• Customers
• Measures
• Targets

Where Is It Used?

Process mapping is a tool used to make certain that key processes are designed in such a way that they focus on achieving the desired outputs and objectives of the organization. The process approach of ISO 9001:2015 includes the requirement to identify and manage the processes of the QMS. This includes focusing on the processes that are most important to achieving the organization's quality objectives. These key processes should not only be monitored by normal activities such as supervision and audits but should also be measured.

How Is It Done?

For each key process, perform the following steps.

Define Outputs and Inputs

- Identify customers and outputs of the process

- Identify the key measures of outputs and understand how they align with the organization's quality objectives

- Identify the suppliers to the process

- Identify the key measures of inputs

- For each measure of input or output:

  - Determine current performance

  - Determine the goal or target

Flowchart the Process

Use the flowchart tool (see tool 1) to show the process as it is. Do not make changes. It is very important to understand the current process in order to understand the causes of gaps between the target and current results. Add the inputs, outputs, and measures to the flowchart. Show which function or department is responsible for each step in the process.

Analyze the Flowchart

Once an accurate as-is representation of the process is created, study the flowchart to determine areas to consider for improvement. Focus on changes that will improve the process performance relative to targets and quality objectives. Examples of things to look for include:

• *Obvious process disconnects* such as outputs that go to the wrong place, obvious missing processing steps, or obvious redundancies. For example, work instructions are routinely misrouted because the distribution list has an error. Consider actions to address issues, for example, changing the routing of misdirected outputs, introducing needed new steps, eliminating redundant steps, and reallocating resources. In this example the only actions needed may be to correct the distribution list and provide a means to keep it up to date.

• *The process has individual activities with problems*, so the overall process does not meet targets, is not effective in meeting quality objectives, or is not efficient, for example, an assembly process where installation of one component is difficult. In such a circumstance, one could identify problems and root causes, develop and implement actions to correct the causes, and measure improvements. If necessary, gather, record, and analyze data on these activities. In this example you may need to analyze the variation of the mating parts to understand how to eliminate the assembly difficulty.

• *The process fails to meet target, but results are stable with no clear problems.* The process is stable and most activities seem OK, but the results do not meet the target, or the process has never met expectations. For example, the production of an engine part may have a cycle time of several weeks due to the time required to correct defects at many stages in the process. If the process exhibits characteristics such as these, consider reengineering the process using a "clean sheet" approach with out-of-the-box thinking or the application of new technology. In this example, a capital project may be required to apply new production technologies that reduce defects to near zero and cut cycle time to a few hours.

• *Opportunities related to risk assessment.* In assessing risks and opportunities one could ask, "what could go wrong?" and apply preventive measures as appropriate. For example, look for opportunities to prevent major equipment failures that could cause costly defects or downtime.

Revise the Process Map, Implement, and Audit

The next step is to revise the map and its flowchart. Ensure that those involved in the process are aware of the changes. Audit the process to ensure ongoing implementation of the changes.

Cautions

Be careful to identify the key processes when constructing a process map and to focus the most effort on improving those areas that are important to meeting your objectives. Flexibility is also important; remember that processes that were unimportant yesterday may be critical to success tomorrow.

TOOL 3: BRAINSTORMING

What Is It?

Brainstorming is a disciplined process used with small groups of people to generate ideas. It is often combined with other techniques to accomplish limited analysis of the ideas generated and to achieve group consensus. One extension of the brainstorming concept is often called the *nominal group technique*, which was developed in the 1960s by André P. J. Delbecq and Andrew H. Van de Ven.

Where Is It Used?

The technique is used in small groups when it is desired to generate a number of ideas that can be used to achieve the group's objectives. Some examples of situations where a group may wish to use the technique include:

- Determining and prioritizing objectives

- Defining potential problems when looking for preventive action opportunities

- Identifying potential causes of known or potential problems

- Defining potential corrective or preventive actions

How Is It Done?

The technique is often performed with the aid of a facilitator to keep the group on track and to ensure that the discipline of the process is maintained. It may be considered to have five phases:

defining the issue, listing alternatives, sorting and combining, clarifying, and developing consensus on priorities.

Phase	Activities
Defining the issue	1. Prepare a clear written description of the issue.
Listing alternatives	1. Each group member lists silently the alternatives he or she can think of.
	2. Go around the group and list one item from each participant. Continue until all items are listed. List each item without challenge.
Sorting and combining	1. Combine items where appropriate.
	2. If the number of items is large, sort them into logical groups.
Clarifying items	1. Discuss the items and ensure each is clear.
	2. Write a concise description of each item or group of items.
Developing consensus	1. Each group member ranks the items.
	2. Rankings are tabulated.
	3. Results are summarized and reviewed by the group with emphasis on items where there is strong consensus.

Cautions

It is important that the objective of the session be clear to all participants. Lack of clarity may cause the process to get off subject. During the development of alternatives, each member of the group must be permitted to list items without discussion or challenge. Discussion needs to be deferred until the clarification phase. The most timidly presented idea may well be the most important.

TOOL 4: GANTT CHART

What Is It?

A *Gantt chart* is a tool for scheduling a series of tasks or events. It lists tasks to be completed on the vertical axis and time on the horizontal axis. It is a tool for clearly illustrating project or program elements or tasks and the amount of time anticipated for completing each. Gantt charts are named for Henry Gantt, an American engineer and social scientist, who developed this tool for use in production management.

Where Is It Used?

Gantt charts are used in many places in the development, deployment, and operation of a QMS. Some examples include:

- Planning the transition to ISO 9001:2015

- Monitoring and controlling an internal audit program

- Structuring an improvement project

- Managing corrective action projects

- Coordinating and tracking the activities of a design and development project

How Is It Done?

A Gantt chart is constructed with a horizontal axis representing time and a vertical axis showing the tasks or events or activities associated with the project or program. Horizontal bars are drawn beside each task (event/activity), representing the sequence and time span for each task. The spans of the bars may overlap if one or more tasks are expected to occur at the same time. Some Gantt charts become elaborate via adding secondary bars and arrows to denote progress, partial completion, and reporting dates. The figure is an example of a Gantt chart for managing a project to achieve compliance with ISO 9001.

Gantt chart to achieve ISO 9001 compliance

Responsibility	Tasks to be completed		Months											
			1	2	3	4	5	6	7	8	9	10	11	12
Top management team	1	Define/clarify organizational responsibilities and interfaces	▓	▓										
Top management team	2	Define objectives and quality policy		▓										
Implementation team	3	Conduct gap analysis against ISO 9001:2015			▓									
Implementation team	4	Identify processes of the QMS			▓									
Implementation team	5	Map processes and draft necessary procedures			▓	▓	▓							
Process facilitator	6	Circulate draft copy of procedures for review/comment						▓						
Implementation team	7	Finalize procedures after staff agreement							▓					
Process facilitator	8	Release ISO 9001:2015–compliant QA documentation								▓				

Continued

Gantt chart to achieve ISO 9001 compliance

Responsibility		Tasks to be completed	Months												
			1	2	3	4	5	6	7	8	9	10	11	12
As assigned	9	Implement system throughout organization		■	■	■								
QA manager	10	Internal audit of implementation effectiveness							■	■				
Implementation team	11	Corrective action including finalization of QA documentation								■	■			
QA manager	12	Pre-audit										■	■	
Registrar	13	ISO certification audit												■

Cautions

Gantt charts do not indicate task interdependencies. One cannot tell from a Gantt chart the impact of a delay in one task on other tasks. To address such issues, tools such as program evaluation and review technique (PERT) charts or the critical path method (CPM) could be used.

TOOL 5: RUN OR TREND CHART

What Is It?

A *run* or *trend chart* is a graphical method for displaying data to show changes over time.

Where Is It Used?

Run charts are often used to track trends in process yields or defects. They are used any time it is useful to visualize trends over time. Some common uses include:

- Trending percent defective or rejected

- Trending parts per million defective or rejected

- Trending capability of a process

- Trending supplier performance

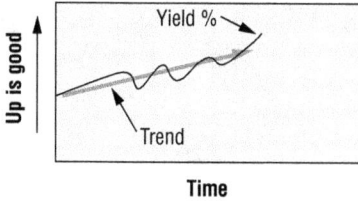

Time

Run chart for process yield

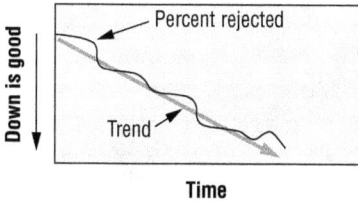

Time

Run chart for percent rejected

How Is It Done?

Any single set of data may be charted over time.

- Select the time period (hour, day, week, month, and so on). This is the horizontal axis of your chart.

- Select a scale for the vertical axis.

- Collect the data for each time period and plot.

Run chart for diameters of parts in production sequence

Diameters of part no. 12-734

Run chart

No. of occurrences	Measured diameter
0	17.384
0	17.383
1	17.382
3	17.381
4	17.38
7	17.379
10	17.378
8	17.377
6	17.376
5	17.375
4	17.374
2	17.373
0	17.372
0	17.371

Run charts are easy to prepare using the graphical tools of most spreadsheet software programs. Run charts are often combined with Pareto charts and histograms for analytical purposes. For example, a run chart may be used to show a trend in percent defective for a product, and an associated Pareto chart may show the defect types. The data from a run chart of product measurement data may be analyzed using a histogram. The same data from the example of the run chart for diameters of a part are used in the illustration of the histogram tool.

Cautions

Be careful in selecting the time period and vertical scales. Too small a vertical scale can give the appearance of big changes in otherwise stable results. Too large a vertical scale can hide significant changes. If the time interval is selected incorrectly, similar distortions may occur.

TOOL 6: HISTOGRAM

What Is It?

A *histogram* is a graphical display of the pattern of variation of a set of data.

Where Is It Used?

Histograms can be used to develop theories about a process. They can also be used to determine whether improvements in process performance have occurred after corrective action was implemented. They can be excellent tools for performing root cause analysis.

Histogram

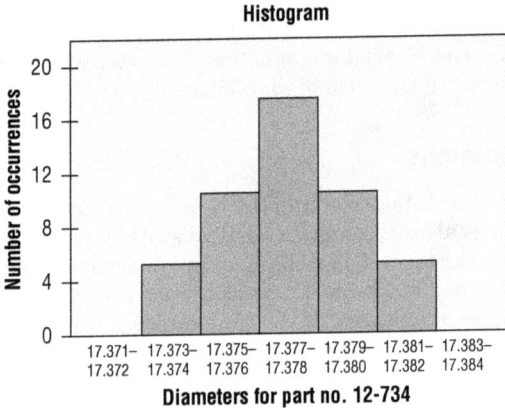

Diameters for part no. **12-734**

How Is It Done?

Assuming a set of data exists, a histogram can be constructed according to the following general methodology:

- Determine the number of data points in the set of data.

- Determine the range of the data (the highest value in the set minus the lowest value).

- Construct a worksheet to tally the data, marking an X for each occurrence of each value (the example shown uses the data from the run chart tool example).

- Decide on the number of cells or classes of data you desire in the histogram; Joseph M. Juran developed guidelines for cells versus data points as follows:

  - 40–50: 6 cells

  - 51–100: 7 cells

  - 101–200: 8 cells

  - 201–500: 9 cells

  - 501 and up: 10 cells

- Determine the size or width of each cell by dividing the range by the number of cells and rounding to a convenient number; be sure that every data point falls into only one cell.

Worksheet

Histogram worksheet

Diameter	Tolerance	No. of occurrences	Cell	Number of occurrences
			17.383–17.384	X
	17.382	4		X X X X
		11	17.381–17.382	X X X X X X X
				X X X X X X X X
		18	17.379–17.380	X X X X X X X X X X
	17.377		17.377–17.378	X X X X X X X X X X X X
		11		X X X X X X X X X
			17.375–17.376	X X X X X X
		6	17.373–17.374	X X X
			17.371–17.372	

- Mark the cell boundaries on the worksheet, starting with a cell that will include the smallest data point and finishing with a cell that will include the largest data point.

- Determine the number of data points in each cell by counting the number of X marks on your worksheet.

- Plot the data on graph paper by indicating the cell width on one axis and the number of data points in each cell on the other axis.

- Label the horizontal and vertical axes, and place a title on the histogram.

Cautions

Be sure that the sample size is adequate to characterize the process (that is, do not make decisions on a set of data that is too small) and that the data points are representative of steady-state process performance. Also, become familiar with the patterns of variation that can be observed in histograms (normal, skewed, truncated, bimodal, and so on) and the implications of such patterns. It is also generally true that histograms will not detect small differences in variability.

TOOL 7: PARETO CHART

What Is It?

The *Pareto chart* is used to show the frequency of occurrence of related sets of data.

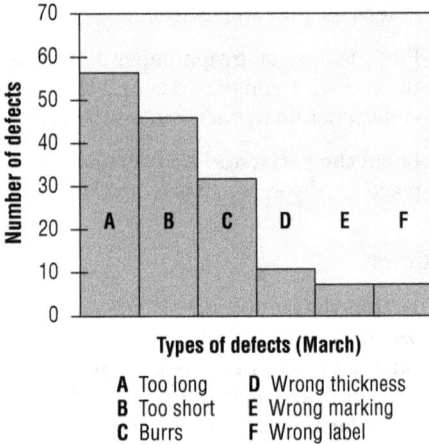

Types of defects (March)

A Too long   **D** Wrong thickness
B Too short   **E** Wrong marking
C Burrs   **F** Wrong label

Where Is It Used?

Pareto charts are often used for analytical purposes to identify the most frequently occurring defects or the most important opportunities for improvement. They are also useful in displaying data so that others can understand priorities. Examples of uses for the Pareto chart include:

- Analysis of the types of defects found in a product over a given period of time

- Analysis of potential causes of a particular problem

- Analysis of the cost of various problems or defect types

- Prioritization of opportunities to improve a process

How Is It Done?

The Pareto chart is based on the idea that in many cases a small number of causes account for a large fraction of nonconformities. To prepare the chart, simply define the appropriate categories (for example, types of defects) and count the number of each for the period under consideration. Plot the data for each category in order of frequency. Select a scale for the vertical axis of the graph that will best display the data.

Cautions

In using the data to make decisions, remember that a Pareto chart of defect types that gives frequency of occurrence does not show criticality or cost. If cost or criticality is more important than frequency of occurrence, a second Pareto chart with, for example, the cost data can be prepared to bring that perspective to the decision-making process.

TOOL 8: FAILURE MODE AND EFFECTS ANALYSIS

What Is It?

Failure mode and effects analysis (FMEA) is a technique for studying the causes and effects of failures before they occur. There are two common variations of FMEA: product FMEA and process FMEA. It is also very common to include not only an analysis of the potential failure modes and their effects but also the criticality of potential failure modes. When criticality is included, the process is normally termed FMECA—or *failure mode, effects, and criticality analysis.*

Where Is It Used?

This technique is usually performed during product design and development and during process development. Since FMEA is one of the most important tools for preventing failures from occurring, its use should be considered when addressing risks and opportunities.

How Is It Done?

Typically, during the design phase of a product development project a designer or a team examines the product or system or subassembly being designed and considers all the ways that failure could occur. Schematics and block diagrams are frequently used. During development of new processes, process engineers often use a similar thought process. In either case, each potential failure is listed in an FMEA table and analyzed as follows:

- For each potential failure, the possible failure modes are listed.

- For each failure mode, a description is developed for each potential effect that such a failure could have.

- For each potential failure, an estimate (on a scale of 1 to 10, with 10 being the worst) is made of severity, probability (or frequency) of occurrence, and detectability (ability to detect the potential cause and prevent the failure).

- The rankings are multiplied to give a risk priority number (RPN) that can be used to prioritize preventive actions.

Usually, an evaluation is made of the potential failure modes, and actions are considered to

Home thermostat FMEA

Part/process	Function of part/process	Potential failure mode	Effect	Cause(s) of potential failure	Severity	Probability of occurrence	Detectability	Risk priority	Action to be taken
Thermocouple (product)	Determines temperature	Loses continuity (open circuit)	System continually calls for heat	Corrosion	9	3	9	243	High temperature cutoff
Battery (product)	Provides power	Loss of power	Shuts off heat	Degrades	9	9	9	729	Weak battery alert signal
Contacts (product)	Closes/opens heating circuit	Fails to close or open circuit	Erratic temperature reading	Corrosion	5	5	5	200	Gold plate contacts
Wall screws (product)	Holds thermostat on wall	Falls off wall	Thermostat falls off wall	Corrosion	1	1	1	9	None

prevent occurrence of or minimize the impact of potential failures with the highest priority.

The example is of an FMEA prepared during the design of a home thermostat.

Cautions

FMEAs can become very cumbersome if every single component of large systems or subsystems is considered. Care must be taken to control the scope of the FMEA, while retaining its integrity. Also, information from an FMEA can often be useful to other activities in an organization. Product FMEA outputs should be shared with organizations performing product safety, maintainability, and serviceability related duties. Process FMEA results should be shared with organizations that will operate and maintain the processes.

TOOL 9: RELIABILITY ANALYSIS

What Is It?

Reliability analysis is the study of how a component, subassembly, assembly, product, or system will perform under stated conditions over time.

Where Is It Used?

Reliability analysis is used primarily in the design phase of the product life cycle and in the analysis of the failure of products shipped to customers. It can also be used as a tool to improve product performance after product release, as an element of supplier selection, and as a tool to measure product performance during manufacture (for example, environmental testing of products at the extremes of specified temperature prior to shipment).

How Is It Done?

There are many excellent books that address the field of reliability analysis. The tools and techniques are many and varied for analysis of both electronic and mechanical subassemblies, assemblies, products, or systems. Reliability analysis is usually conducted by reliability engineers specially trained in this field. The purpose of this description of reliability analysis is to direct the reader to one of the many reference sources for reliability (see, for example, O'Connor and Kleyner, *Practical Reliability Engineering*, 5th ed., Wiley & Sons, 2012, or Martin L. Shooman, *Probabilistic Reliability—An Engineering Approach*, McGraw-Hill, New York, 1965) or to recommend that a reliability engineer be involved in analysis of problems or issues related to the performance of products over time.

It is also worth considering expanding the skills of existing QA staff by encouraging the study of the body of knowledge for the Certified Reliability Engineer certification offered by ASQ.

Cautions

Reliability analysis is certainly a science, but there are many assumptions that typically must be made that can dramatically influence the outcome of any analysis. Assumptions must be made cautiously. Also, reliability decisions are not made in a vacuum. There are usually cost and performance trade-offs to be made, and such trade-offs need to be evaluated in the context of the objectives of the organization, considering the inputs from all concerned functions, including manufacturing, sales, service, marketing, finance, test, and purchasing, in addition to engineering and quality.

TOOL 10: IDENTIFYING EXTERNAL AND INTERNAL ISSUES—USE OF SWOT ANALYSIS

What Is It and How is it Done?

Clauses 4 and 6 of ISO 9001:2015 have introduced a few requirements that may be new to many users of this standard. How can an

organization proceed to address the requirements related to identifying and determining
external and internal issues relevant to the organization and identifying interested parties and
determining their requirements?

There are several ways to address these
requirements, and many organizations already
have processes established and deployed as elements of a formally implemented strategic planning process. If your organization does not have
processes already in place, you can consider the
following approach to initiating the activities of
your organization that address requirements for
both determining external and internal issues
relevant to the organization and identifying
interested parties and their requirements.

Figure 5.1 shows a model of the activities of
the strategic and tactical planning processes
typically employed by organizations.

In the upper-left corner are activities to be
that can be employed to identify both external
and internal issues relevant to the organization and to identify interested parties and their
requirements. The organization can consider
completing a matrix related to the requirements
of clauses 4 and 6 of ISO 9001:2015, contemplating the strengths, weaknesses, opportunities and threats (SWOTs) related to each of the
requirements. Figure 5.2 shows examples of how

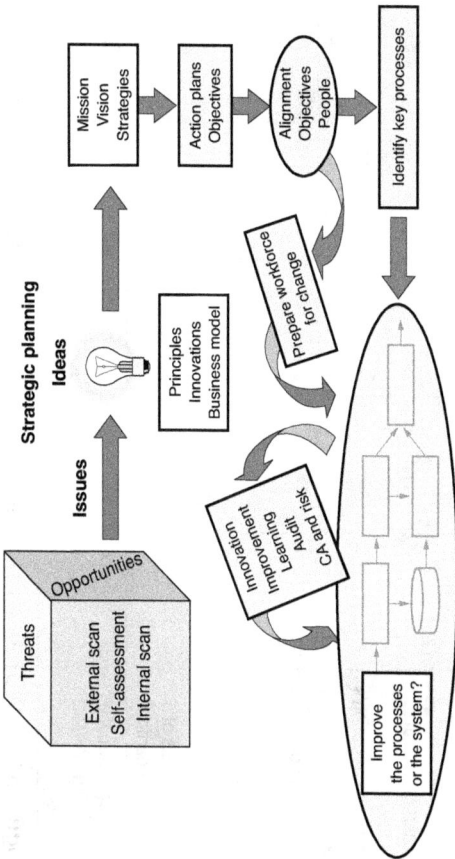

Figure 5.1 Example of a model for strategic and tactical planning processes.

ISO requirement	Strengths	Weaknesses	Opportunities	Threats	Process actions
Determine external/internal issues relevant to the organization's purpose					
Monitor information regarding issues					
Determine interested parties					
Requirements of interested parties					
Review information regarding interested parties					

Figure 5.2 SWOT analysis format for considering requirements for clauses 4 and 6.

a few of the requirements can be extracted and a SWOT analysis performed.

After contemplating SWOT for each requirement, the organization can then decide what processes, if any, it should define and deploy to address the requirements and the SWOTs it identified that merit action.

Retention of documented information on such activities will provide objective evidence of consideration of the requirements and the actions taken to ensure compliance.

Where Is It Used?

Such an approach would be most beneficial when it is performed during the planning phase of the development of the QMS.

Cautions

Strategic planning is an important process that requires top management involvement. Establishing and integrating the QMS requirements into the strategic planning of the organization is one of many elements of the process that must be incorporated. The above approach will provide a form and structure to ensure that key elements are considered for attention. Such an approach is just a starting point, but it will initiate consideration of processes that may be critical to sustainability of the organization.

TOOL 11: PROCESS CAPABILITY STUDIES

What Is It?

A *process capability study* is a statistical technique for determining the behavior of a process, including the people involved, the machines, the materials, and the methods used. Process capability studies can be performed on simple processes—for example, the manufacture of a machined part—or an entire assembly or even on an entire production operation, such as the manufacturing of an automobile, although the common application is on a part, a subassembly, or an assembly. The objective of a process capability study is to determine the natural behavior of a process and to understand what, if anything, needs to be changed to achieve the desired results. For example, if a process capability study indicates that a milling machine is capable of holding a tolerance of ±.003 inches but the specified tolerance is ±.002 inches, then corrective action of some kind is needed (such as changing the tolerance or the machine used or both, or other creative actions).

Where Is It Used?

Process capability studies can be used to address many different kinds of problems in manufacturing, inspection, test, management, or engineering. Such studies can be used just about anywhere it is important to understand the nature and behavior of the distribution of the output of a process.

How Is It Done?

Entire books have been written to describe how to perform process capability studies. Here we can provide only a very general and brief overview. Performing a process capability study typically follows the scientific method: (1) conduct an experiment to gather data from the process, (2) make a hypothesis about the process from the data gathered (for example, that the data exhibit only natural variation), (3) test the hypothesis by using statistical techniques to interpret the data (for example, plot the data on a control chart) to understand and separate normal variation and special cause variation, (4) change the process, if necessary, based on the interpretation of data, and continue performing experiments until the process is operating in a way that consistently produces parts within specification at an acceptable cost.

This treatment of process capability studies is intended only to expose the fact that there are a variety of techniques available to perform capability studies of processes that can be powerful tools in achieving conformance to requirements, customer satisfaction, and cost control. Further study would be necessary to effectively apply such techniques in a manufacturing environment.

Cautions

Gathering and interpreting the data obtained when conducting a process capability study can be tricky: care needs to be taken in measuring, the data collected must be representative of the process performance (for example, different shifts, different operators), control chart construction can be tedious, and patterns of variation can be difficult to discern. If one is new to the art and science of conducting process capability studies, it is advisable to solicit advice and counsel from an experienced practitioner in the organization, who can typically be found in QA or engineering.

TOOL 12: CAUSE-AND-EFFECT DIAGRAM (FISHBONE/ISHIKAWA DIAGRAM)

What Is It?

A *cause-and-effect diagram* is a tool that can be used to show the relationship between a characteristic or an output of a process and its potential cause factors. The cause factors are organized into categories and displayed on a diagram. The purpose of the diagram is to facilitate broad thinking about a process and how to control the significant cause factors so that the desired effect or result is obtained.

The cause-and-effect diagram was created by Dr. Kaoru Ishikawa and is sometimes called an *Ishikawa diagram*. It is also called a *fishbone diagram* since, when it is drawn, it resembles the skeleton of a fish.

Where Is It Used?

Cause-and-effect diagrams are frequently used in problem solving. They can be used anywhere it is desired to understand the cause factors or characteristics that influence the outcome of a process. They are commonly used when performing corrective action, when a process is producing out-of-specification product, when trying to understand the root cause of a customer complaint, or when refining a manufacturing process to improve throughput.

How Is It Done?

- Step 1: Clearly identify the effect or characteristic of the process you desire to study, or in other words "what is the problem you desire to analyze?"

- Step 2: Identify the major cause factors that contribute to or influence the effect; in a "classic" cause-and-effect diagram for manufacturing processes, the major cause factors are categorized as:

  - Materials

  - Machines

  - Measurements

  - Methods

  - Man (that is, the people working in the process)

- Step 3: Construct the diagram by positioning the effect in a box on the right side of a piece of paper or on a computer screen and drawing a horizontal arrow from the left to the box. Then insert branches above and below the horizontal arrow that feed into the horizontal arrow (see figure) to indicate significant cause factors in the categories indicated in step 2.

- Step 4: Develop the diagram by thinking through the significant cause factors for each category in step 2 and adding or modifying the cause factors.

- Step 5: Discuss the possible cause factors and decide what actions are appropriate to improve process performance.

- Step 6: Implement appropriate corrective or preventive actions and perform follow-up evaluations to ensure that results are compatible with expectations.

- Step 7: Institutionalize process improvements via documentation, training, and audit. See the figure for an example of a fishbone diagram created to understand the root cause of a milling machined part being out of spec.

Cautions

The quality of the analysis of causes is only as good as the thinking of the individuals doing the

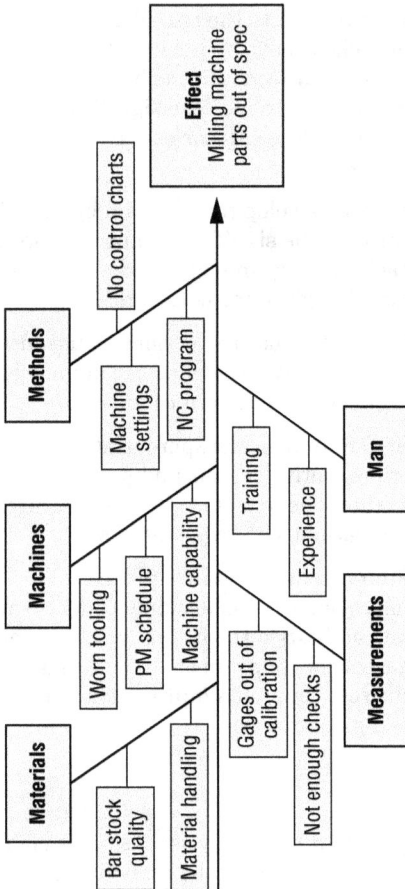

Effect
Milling machine parts out of spec

Methods
- No control charts
- Machine settings
- NC program

Machines
- Worn tooling
- PM schedule
- Machine capability

Materials
- Bar stock quality
- Material handling

Man
- Training
- Experience

Measurements
- Gages out of calibration
- Not enough checks

analysis, so it is important to involve all the individuals who can contribute to the identification of potential causes. In the milling machine example, it may be appropriate to have manufacturing engineers, machine operators, quality assurance engineers, purchasing agents, materials engineers, and production supervisors involved in an analysis of possible causes of the problem.

TOOL 13: PROBLEM SOLVING

What Is It?

Defining and solving problems is a key component of continual improvement. A problem can be broadly defined as "any undesired state." With regard to ISO 9001:2015 requirements, problem solving is a critical tool for achieving effective corrective action and continual improvement.

Where Is It Used?

Problem-solving techniques are generally appropriate:

- To correct causes of nonconformities in product

- To address and correct causes of customer complaints

- To address and correct situations that dissatisfy customers

- To improve processes, even processes that are meeting requirements

How Is It Done?

Perhaps the most important step in problem solving is to address the correct problems. You should first prioritize the problems you face (Pareto charts may help). Once a high-priority problem has been identified, use a disciplined approach to solve it. The problem-solving concept can be described in many ways. It can be generalized as shown in the following table:

Analysis	• Identify and define the problem
	• Develop measurements or obtain data to understand the current situation and for later use in determining the extent to which the problem has been solved (trend charts may be of use in tracking progress)
	• Perform analysis to identify causes of the problem: tools such as cause-and-effect diagrams, scatter plots, and control charts may be needed.

Continued

Analysis *(continued)*	• Prioritize potential causes—problems often have more than one causative factor (Pareto analysis may be a help)
	• Develop potential solutions to address the important cause(s)
	• Develop and analyze potential solutions and select the one(s) most likely to solve or reduce the problem
Action	• Implement the solution(s)
	• Implementing solutions means change—manage the change properly
	• If the change or its impact is large, consider a pilot implementation
Measure results	• Consider the feasibility of determining if you can "turn the problem off and back on" by alternately applying and withdrawing the solution(s)
	• Measure the results to determine how much progress has been made
Institutionalize	• If the problem has been solved or sufficient progress made, take action to ensure the solution is permanent:
	– Change the documentation
	– Train personnel
	– Audit to ensure ongoing effectiveness of the solution
	• If the problem is still significant, return to analysis

Cautions

Problems come in many forms, and it may be appropriate to use tools and problem-solving sequences different from those described here. Organizations should consider providing training and facilitation for the process. The first and last steps are often the most important but the least often done well. Following these steps, clearly defining the important problems, and institutionalizing the changes required for effective solutions are critical elements of effective problem solving.

TOOL 14: HOW TO CONDUCT AN IMPROVEMENT PROJECT

What Is It?

When an organization is implementing a QMS in conformity with ISO 9001:2015, it is difficult not to be directed to improvement of the organization's processes, products, and services. In spite of the continual need for improvement, it is not uncommon for organizations to have no defined methodology for addressing improvement opportunities.

There are several forms of improvement an organization should consider, ranging from systemwide improvement of the overall QMS or any of its processes to local improvement of individual processes that are producing results that do not meet requirements.

This tool is focused on providing a simple model to guide the improvement of individual processes for which the results are not meeting requirements or expectations. This tool is intended for use by anyone in an organization.

Where Is It Used?

Improvement projects can be initiated in virtually any area of an organization. Typical examples include projects to:

- Eliminate waste

- Reduce defectives in a manufacturing process

- Eliminate billing errors

- Reduce customer complaints

- Increase customer service rate

- Improve acceptance rate of purchased material

How Is It Done?

This tool requires no special training other than a willingness to invest energy into addressing each of the seven simple activities described below. It provides a structure for addressing a wide variety of conditions where improvement is needed. It also encourages pursuit of *meaningful corrective action rather than reactive correction* of undesirable conditions. This model provides an opportunity to use many of the other tools presented in this chapter. It is not the only model that an organization can adopt to guide improvement initiatives, but if no other model or procedure exists, this is a good starting point.

The activities we propose are as follows:

- Activity 1: Determine project charter or project definition (that is, what you are going to do)

- Activity 2: Describe current state

- Activity 3: Determine root causes

- Activity 4: Consider possible solution options and select the most promising for implementation

- Activity 5: Implement solutions

- Activity 6: Evaluate solutions

- Activity 7: Standardize solutions

```
┌─────────┐   ┌─────────┐   ┌─────────┐   ┌─────────┐
│ Charter │──▶│ Current │──▶│  Root   │──▶│ Develop │───┐
│         │   │  state  │   │ causes  │   │solutions│   │
└─────────┘   └─────────┘   └─────────┘   └─────────┘   │
                                                        │
   ┌────────────────────────────────────────────────────┘
   │
   ▼
┌──────────┐   ┌──────────┐   ┌────────────┐
│Implement │──▶│ Evaluate │──▶│Standardize │
│solutions │   │solutions │   │ solutions  │
└──────────┘   └──────────┘   └────────────┘
```

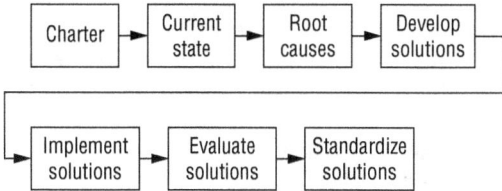

Let us consider application of this model in an example:

> *Situation: The placement staff of a personnel agency has a goal to achieve 20 job placements each week for its clients. For the last several months, the agency staff has been averaging 14 job placements each week. Management wants to improve placement performance.*

The model can be applied to this issue as follows:

- *Activity 1: Determine project charter.* This could be as simple as stating that an improvement team will identify and deploy actions to increase the weekly placement average to 20 from the current level of 14.

- *Activity 2: Describe current state.* Currently, performance has been steady at 14 per week for the past several months, and the agreed goal is 20 placements per week.

- *Activity 3: Determine root causes.* After analysis of placement data for the past six months, the improvement team agreed that there were three primary issues negatively impacting placement rate—issue A, issue B, and issue C.

- *Activity 4: Develop solutions.* After brainstorming many possible actions to address the root causes of the lower-than-goal level of placements, it was decided to implement three new processes to attempt to improve placement performance—process E, process F, and process G, along with a measurement system to track the effectiveness of the newly implemented process.

- *Activity 5: Implement solutions.* Introduction of new or modified processes should include consideration of appropriate documentation and training to ensure competent and consistent implementation.

- *Activity 6: Evaluate solutions.* After three months of implementation, the effectiveness of the newly implemented processes is evaluated, with measurement of effectiveness accomplished by review of the graphs of actual placements compared with the goal and the performance during the preceding time period. In addition to graphs of the overall group performance, the improvement team can review the performance of each individual.

- *Activity 7: Standardize solutions.* After evaluation of the effectiveness of the newly implemented processes, the processes that have had a favorable impact are "institutionalized" by providing the documentation and training needed to ensure consistent implementation in the organization.

This is a simplified example of how a systematic approach to improvement can be applied to many of the product or service delivery–related processes in an organization.

Cautions

When organizations encounter processes that are not yielding expected results, it is not uncommon to leap into attempting to correct the perceived issues without understanding whether the process is capable of meeting requirements or determining the root causes of deviation from requirements or expectations. Such correction attempts may yield a "quick fix" but typically do not result in sustainable improvement. The approach described here is designed to yield sustainable improvement.

Index

Ask a Librarian

Did you know?

- The ASQ Quality Information Center contains a wealth of knowledge and information available to ASQ members and non-members

- A librarian is available to answer research requests using ASQ's ever-expanding library of relevant, credible quality resources, including journals, conference proceedings, case studies and Quality Press publications

- ASQ members receive free internal information searches and reduced rates for article purchases

- You can also contact the Quality Information Center to request permission to reuse or reprint ASQ copyrighted material, including journal articles and book excerpts

- For more information or to submit a question, visit **http://asq.org/knowledge-center/ask-a-librarian-index**

Visit www.asq.org/qic for more information.

TRAINING   CERTIFICATION   CONFERENCES   MEMBERSHIP   **PUBLICATIONS**

ASQ
The Global Voice of Quality®

Belong to the Quality Community!

Established in 1946, ASQ is a global community of quality experts in all fields and industries. ASQ is dedicated to the promotion and advancement of quality tools, principles, and practices in the workplace and in the community.

The Society also serves as an advocate for quality. Its members have informed and advised the U.S. Congress, government agencies, state legislatures, and other groups and individuals worldwide on quality-related topics.

Vision

By making quality a global priority, an organizational imperative, and a personal ethic, ASQ becomes the community of choice for everyone who seeks quality technology, concepts, or tools to improve themselves and their world.

ASQ is...

- More than 90,000 individuals and 700 companies in more than 100 countries

- The world's largest organization dedicated to promoting quality

- A community of professionals striving to bring quality to their work and their lives

- The administrator of the Malcolm Baldrige National Quality Award

- A supporter of quality in all sectors including manufacturing, service, healthcare, government, and education

- YOU

Visit www.asq.org for more information.

TRAINING   CERTIFICATION   CONFERENCES   MEMBERSHIP   **PUBLICATIONS**

ASQ
The Global Voice of Quality®

ASQ Membership

Research shows that people who join associations experience increased job satisfaction, earn more, and are generally happier*. ASQ membership can help you achieve this while providing the tools you need to be successful in your industry and to distinguish yourself from your competition. So why wouldn't you want to be a part of ASQ?

Networking

Have the opportunity to meet, communicate, and collaborate with your peers within the quality community through conferences and local ASQ section meetings, ASQ forums or divisions, ASQ Communities of Quality discussion boards, and more.

Professional Development

Access a wide variety of professional development tools such as books, training, and certifications at a discounted price. Also, ASQ certifications and the ASQ Career Center help enhance your quality knowledge and take your career to the next level.

Solutions

Find answers to all your quality problems, big and small, with ASQ's Knowledge Center, mentoring program, various e-newsletters, *Quality Progress* magazine, and industry-specific products.

Access to Information

Learn classic and current quality principles and theories in ASQ's Quality Information Center (QIC), *ASQ Weekly* e-newsletter, and product offerings.

Advocacy Programs

ASQ helps create a better community, government, and world through initiatives that include social responsibility, Washington advocacy, and Community Good Works.

Visit www.asq.org/membership for more information on ASQ membership.

*2008, The William E. Smith Institute for Association Research

TRAINING   CERTIFICATION   CONFERENCES   **MEMBERSHIP**   **PUBLICATIONS**

ASQ
The Global Voice of Quality®